DATE DUE

Sca   y
T

DISCARDED

Demco, Inc. 38-293

D1502265

HEINEMANN
Portsmouth, NH

NOV 15 2010

Heinemann
361 Hanover Street
Portsmouth, NH 03801–3912
www.heinemann.com

Offices and agents throughout the world

© 2008 by Michael Klentschy and Laurie Thompson

Library of Congress Cataloging-in-Publication Data
Klentschy, Michael.
  Scaffolding science inquiry through lesson design / Michael Klentschy, Laurie Thompson.
      p. cm.
  Includes bibliographical references.
  ISBN-13: 978-0-325-01154-7
  ISBN-10: 0-325-01154-0
  1. Science—Study and teaching (Elementary)—Curricula—United States.
2. Science—Study and teaching (Middle school)—Curricula—United States.
3. Lesson planning—United States. 4. Inquiry-based learning—United States.   I. Thompson, Laurie.   II. Title.

LB1585.3.K59 2008
372.35'043—dc22   2008037398

Editor: Robin Manning Najar
Development editor: Alan Huisman
Production: Vicki Kasabian
Cover design: Shawn Girsberger
Typesetter: ICC Macmillan Inc.
Manufacturing: Louise Richardson

Printed in the United States of America on acid-free paper
12 11 10 09 PAH 2 3 4 5

Contents

This book is dedicated to all of the staff from Project SEED and CAPSI in Pasadena, California, where the genesis of the scaffolded guided inquiry took place, and to Elizabeth, Mercy, Yolanda, and Yvonne from the Valle's Imperial Project in Science for their leadership and hard work in making the ideas become practice with the teachers and children in Imperial County, California.

Introduction

Classroom teachers, school principals, and school districts across the United States are facing an enormous challenge. Some believe that the national movement toward standards, assessment, and accountability emphasizing reading, writing, and mathematics, as measured by high-stakes standardized tests, threatens progress in science education (Jorgenson and Vanosdall 2002). School districts, under pressure to improve test scores in basic skills, allot the majority of the instructional day to reading, writing, and mathematics.

It is vital that teachers design classroom science instruction that students learn from and find meaningful. Often the outcomes of daily classroom instruction are not fully realized until several years later. There are lots of stories about students from backgrounds not usually associated with success in school who, because of an inspirational and motivational elementary school teacher, do well in high school and college and ultimately have rewarding careers in math, science, or engineering.

As a case in point, six years ago, I received the following letter.

Dear Dr. Klentschy,

You may not remember me, but my name is Rosa Lopez. I met you at the parent institute two years ago. You were very busy cooking pancakes for us then, but I wanted to write you about my daughter, Maria. She is leaving Kennedy Middle School and will be attending Southwest High School next year. Her grandmother and I are very proud of Maria. She will be the first person in our family to attend high school.

Four years ago I noticed a real difference in my daughter. She became very excited about science at McKinley Elementary School. She would come home every day and would talk about science around the dinner table with me and her grandmother. We learned about her science lessons with electricity, magnets, plants, and motors. She even wrote to me in Salinas [California] when I was there picking lettuce. Maria now wants to be a scientist or a doctor. Her grades are all very good now. I just received her test scores in the mail and she is in the 90s in all of her subjects.

I wish to thank you and her teachers for the wonderful science program you started four years ago. It helped my daughter find a pathway for her life. One that will be brighter than mine, but ever so valuable to our poor community.

God bless you.

Sincerely,
Rosa Lopez

Maria's excitement over her scientific discoveries as a student at McKinley Elementary School, in El Centro, California, ultimately led her to the Stanford University School of Engineering, where she is a student today.

The science program Rosa Lopez mentions in her letter is the Valle Imperial Project in Science (VIPS), in Imperial County, California. Imperial County is a geographically isolated region of southeastern California bordered by Mexico on the south and Arizona on the east. It is one of California's largest counties in terms of area, but it is sparsely populated, and its residents are among the poorest in the state in terms of real income. Agriculture is the primary industry.

The students in Imperial County are predominately Hispanic English language learners, and most of them are eligible for the federal free and reduced-price lunch program. There are fourteen elementary and middle school districts. Six rural districts have only one school each; six districts have between three and six schools each; and there are two larger districts, one with eleven schools, the other with twelve. El Centro is the economic and administrative hub of the county, and the El Centro Elementary School District has nine elementary and two middle schools. El Centro is also the lead VIPS district.

The Valle Imperial project began in 1996 as a countywide collaborative partnership comprising the fourteen school districts; San Diego State University, Imperial Valley Campus; and the California Institute of Technology. It was funded by a grant from the National Science

Foundation. One of the project's goals was to increase the number of high school students enrolling in and successfully completing challenging high school science courses, becoming university science majors, and going on to careers in science, mathematics, engineering, and technology. The project focused on redesigning science lesson plans and instruction in elementary and middle school classrooms, based on the belief that early exposure to a high-quality elementary science program is key to being able to do well in challenging high school science courses and meet university entrance requirements.

The project designers recognized that understanding complex scientific reasoning takes time and that concepts and thinking skills are best embedded within each unit of instruction. This approach is consistent with the National Research Council's recommendations that students best learn complex science content when classroom teachers plan instruction that (1) activates student prior knowledge by linking the strange new (previously unencountered knowledge) to the familiar old (prior knowledge), (2) helps students develop a deep foundation of factual knowledge in the context of big scientific ideas, and (3) employs metacognitive learning approaches that help students "own" the material (National Research Council 1999, 2005).

When students link the familiar old to the strange new in a series of individual lessons related to a standard, they form subconcepts, and these subconcepts, linked together, lead to a conceptual understanding of big ideas in science. Metacognitive approaches to learning ownership (graphic organizers, Venn diagrams, cause-and-effect flowcharts, and so on), long used in literacy instruction to encourage students to think about their own thinking, can also be applied successfully in science instruction.

VIPS staff members began by analyzing the existing state-adopted curriculum to discover how it aligned with the National Science Education Standards (National Research Council 1996) and the state science content standards California was then drafting. (They were implemented in 1998.) They found a poor match. They also found that English language learners could only access the state-adopted curriculum through supplemental workbooks.

They then went on to determine what kind of instruction would best deliver an aligned curriculum. Although the general population and even some educators often think of classroom inquiries as unstructured, open-ended activities, the National Research Council

(2005) maintains that inquiry follows a path from teacher centered to student centered: (1) the teacher demonstrates a science concept; (2) the teacher introduces questions that prompt students to investigate a science concept and processes that allow them to do so (directed inquiry); (3) the teacher introduces a science standard and lesson goals, and the students formulate a question, predict an answer to the question, devise a plan for answering the question, collect data, link claims to evidence, and draw conclusions (guided inquiry); and (4) the students create and investigate their own questions (open/full inquiry). The National Research Council reiterated these views in 2000:

> Investigations can be highly structured by the teacher so that students proceed toward known outcomes, such as discovering regularities in the movement of pendulums. Or investigations can be free-ranging explorations of unexplained phenomena. . . . The form that inquiry takes depends largely on the educational goals for students, and because these goals are diverse, highly structured and more open-ended inquires both have their place in science classrooms. (National Research Council 2000, 10–11)

Given the current era of standards, assessment, and accountability, the project directors felt that guided inquiry would give teachers the opportunity to address the science content standards and at the same time give students opportunities to gain a deeper understanding of the required content and develop complex science reasoning skills.

But we soon realized that guided inquiry is a complex process severely hampered when students lack content knowledge, inquiry experience, and classroom resources. In addition, students are often unable to create meaningful inferences from the data they collect. In light of these limitations, we felt that student understanding could best be achieved through *scaffolded* guided inquiry (with science notebooks and classroom discussion the primary scaffolding tools) presented in predesigned lessons structured around an intended curriculum (standards), an implemented curriculum (what is taught), and an achieved curriculum (what students learn).

This book examines how lessons built around classroom discussion and science notebooks help students avoid misconceptions as they make meaning during inquiry and thus develop a deeper understanding of complex science concepts and the ability to reason scientifically. Thus, two important features were joined to form a powerful instructional tool. A research-based planning template is used to structure and make coherent the design of instruction, and the tools of focused classroom discussion and student science notebooks are used to enhance instructional delivery. Chapters 1 and 2 describe the structure of the research-based planning template and the tools (classroom discussion and science notebooks) necessary to scaffold guided inquiry. Chapters 3–7 focus on planning the lessons using this template for the intended curriculum and presenting those lessons as the implemented curriculum using classroom discussion and science notebook strategies. Examples of student work and specific writing and discussion prompts are included. Chapter 8 focuses on the achieved curriculum and discusses strategies for providing feedback to students that will extend their thinking, correct their misconceptions, and deepen their ability to reason scientifically. The final chapter, Chapter 9, summarizes several studies showing results that have been achieved using this instructional model.

Using a Research-Based Template to Plan Classroom Instruction

1

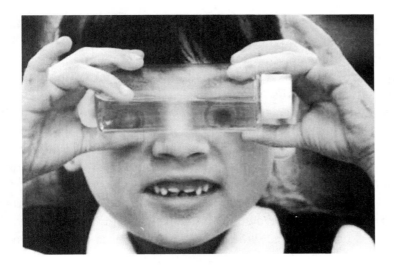

The current national and state focus on standards, assessment, and accountability due to the No Child Left Behind legislation challenges most classroom teachers to utilize instructional planning instruments and lesson design models that provide standard specific instruction and to gather evidence of student proficiency through both student work and testing. Instructional programs emphasizing inquiry are often criticized for being too open ended. The critics of this approach feel that inquiry often lacks a well-defined alignment to specific content standards and does not provide a reliable means of assessment.

As a result of states adopting science content standards to meet federal mandates, it has become an increasing challenge for schools and classroom teachers to develop complex scientific reasoning abilities among the students rather than just a recitation of science facts. To have a real understanding or to be able to make real meaning of the big ideas in science, students need to be able to extend their ability to explain these concepts and their relationship to big ideas in science in their own words and based upon their own experiences. Using an alignment model or a lesson design model to plan guided inquiry that addresses both the need for content understanding and the development of complex scientific reasoning abilities will provide classroom teachers with the structure necessary to accomplish this task.

The designers of the Valle Imperial Project in Science (VIPS) model recognized that development of complex scientific reasoning takes time and that to be effective every science unit should be embedded with a lesson design that addresses the development of science concepts and thinking skills. This thinking was consistent with the recommendations from the National Research Council (1999, 2005). The structure of the VIPS lesson design model and classroom instruction focuses on maximizing student opportunity to develop complex scientific reasoning. The ability of students to actually make meaning or understand the goals of what the intended lesson was trying to achieve are essential to student learning.

The goal of every classroom teacher should be for students to make meaning and develop deep science conceptual and procedural understanding from classroom science experiences. Often this goal is difficult to achieve due to a disconnect between what should be taught, what actually *is* taught, and what students learn in the design of classroom instruction (Marzano 2003). These breakdowns in curriculum alignment and student learning may be a function of poor lesson design and planning or from teachers' "leap of faith" that the science curriculum materials that they use are aligned.

The research-based components of the VIPS lesson design provide teachers with a planning structure and provide practical suggestions to support student success. Research on how students learn science indicates that the development of deep conceptual understanding in science requires time and can be enhanced through providing supports, scaffolds, and prompts that guide students to enhance their scientific reasoning abilities (National Research Council 2005). Thus, classroom teachers must guide the inquiry process in order to develop, in their students, both deep conceptual understanding and the reasoning ability to formulate explanations based upon evidence.

The lesson design model VIPS developed uses a scaffolded guided inquiry approach for lesson planning. Using this approach, classroom teachers can address the need to systematically focus on a sequential set of instructional units all aligned to state content standards over several years to develop practice on the part of students in the use of the scaffolds through a consistent instructional approach. This approach provides classroom teachers with a lesson design that guides inquiry using scaffolds that are designed to place the focus of instruction on the actual intended curriculum through the implemented curriculum and attain the development of student understanding of the science content described within the standards and the development of complex reasoning abilities, such as analyzing and interpreting data or formulating claims from evidence collected during an investigation.

A planning model for scaffolded guided inquiry is a valuable structure for the alignment of the three critical elements of lesson planning—intended, implemented, and achieved curricula. By using a consistent approach to lesson planning and implementation, students are provided with *sameness* or consistency. Because students need to generate their own meaning regarding the science content being learned, the psychological principle of sameness is important. Marzano (2003) states that sameness is critical to the process of learning. When students are presented with processes that are similar through the consistent exposure to writing and discussion scaffolds or prompts, students learn how to do inquiry and develop the ability to make evidence-based explanations from their science investigations. This requires that teachers plan the learning experiences for children with careful thought. These learning experiences also need to be sequenced over several units and years because students need time and practice to learn how to do inquiry. It is therefore important for teachers to have the support of a lesson design model that is based on student success and carefully crafts an alignment between the intended curriculum, the implemented curriculum, and the achieved curriculum and is consistent in its approach to scaffolding.

Figure 1–1 depicts the lesson design model and the built-in scaffolds for guided inquiry that aligns the intended, implemented, and achieved curricula. Each of the three components of this model are introduced next and are discussed in subsequent parts of this book.

Figure 1–1 Aligned Planning Model

Intended Curriculum
Big Ideas—Public Announcement

Lesson Content Goals	**Guiding Questions**
1. ⟷	1. Make public
2. ⟷	2.

State Standard Addressed

Implemented Curriculum **Opportunities to Learn**	**Achieved Curriculum** **Feedback Guide**
Kit inventory	Science notebook
Working word wall—synonyms (tags)	Formative assessment of
Engaging scenario—connect to world	teaching and learning
	Proficiency/guidance for
	improvement

❖ Focus Question ——————→ ❖ Focus Question
  A question that leads to construction
  of knowledge about lesson content goals

❖ Prediction ——————→ ❖ Prediction
  I think or predict that _____ because _____.
  If _____, then _____.

❖ Plan ——————→ ❖ Plan
❖ Data Organizer ❖ Data Organizer
❖ Data ❖ Data
  Plan, organize, set expectations

Making Meaning Conference
• Class graphic organizer (key concept),
  thinking map
• Sharing data, group analysis
• Claims and evidence emerge—identify
  on organizer

❖ Claims and Evidence ——————→ ❖ Claims and Evidence
❖ Conclusions ❖ Conclusions

Closure
• Share, discuss, challenge claims and
  evidence, revisit big ideas ——————→ ❖ Reflection
• Revisit predictions   Support or change
• Next steps, new questions   thinking

Key Elements of the Intended Curriculum

The model begins with what we call the *intended curriculum* or that which is expected to be taught. The intended curriculum identifies the big ideas of science and the specific district, state, or national science content standards that are to be addressed within a given lesson. The intended curriculum also specifically identifies the content goals that the lesson should achieve and the guiding questions that help students focus on the inquiry that will help them achieve the goal of the lesson. In order to be effective, the guiding questions must be aligned with the lesson content goals. This alignment of the lesson content goals with the guiding questions will also provide classroom teachers with an effective means of establishing criteria for student success in the lesson. The alignment defines what students should know and be able to do by the end of the lesson and provides clarity to the planning process for classroom teachers in terms of what should be learned by students as a result of the lesson.

Key Elements of the Implemented Curriculum

The next phase of the lesson design models is the *implemented curriculum*. This is where the meat of the inquiry lies and where much of the activity of investigative science happens. The implemented curriculum has four phases, each designed to maximize student opportunity to learn.

The first phase, Set the Stage for Learning, is the unit or lesson opening that integrates research-based tools such as a kit inventory, working word walls, and vocabulary development. These tools provide a basis for introducing the unit and are the first step for students in developing fluency and contextual use of scientific vocabulary.

The second phase, Formulate Investigable Questions and Predictions, opens with high-interest scenarios that lead to student engagement in the lesson. The engaging scenarios are written in such a way to pose a problem that must be solved through an investigation.

The engaging scenarios may also be serialized throughout a unit of study and may be used to connect lessons. From these high-interest scenarios, students determine "what they want to find out" by formulating focus questions that address the problem posed in the engaging scenario. In this phase, science notebooks and discussion are used to help develop and record students' questions and predictions. This phase lays the groundwork for investigative science on the part of the student.

The third phase of the implemented curriculum is Plan and Conduct the Investigation. Using small-group and class discussion, students devise a plan of action to answer the question created in Phase 1, utilize science notebooks to develop graphic organizers to record their observations and data collection, conduct the investigation, and record their collected data or observations in their science notebooks. This phase is critical in helping students create both a general and specific plan of action that will guide them through the investigation.

The fourth phase, Make Meaning, helps students make claims based upon their gathered evidence through the use of discussion and the entries in their science notebooks. Students are guided in the drawing of conclusions, reflecting on their experiences, and the development of new questions. Each of these phases are specifically planned to integrate the tools of writing and discussion as key components for students to gain meaning from the science lesson.

Key Elements of the Achieved Curriculum

The last element of the lesson design model is the *achieved curriculum*. The achieved curriculum is directly aligned to the intended and implemented curricula. Each of the four phases of the implemented curriculum specifically guides the students to discuss key elements of the investigation and then to make appropriate notebook entries. There is a direct relationship between the implemented curriculum component and the associated notebook entry.

Student science notebooks are indicators of what is achieved in the lesson. By reading the entries, the classroom teacher can determine

Scaffolding Science Inquiry Through Lesson Design

if an alignment was consistent with the intended and implemented curricula. Student science notebooks also serve as tools to reflect individual student learning, class progress, and the effectiveness of the lesson. An important element to measure the degree of attainment in the achieved curriculum is the development of a feedback guide that is specific to the intended and implemented curricula. The student science notebook then becomes an effective formative assessment tool that teachers can use to see if the students are actually making the connection between what teachers were hoping to teach and what they actually taught.

This lesson design model is filled with guided inquiry scaffolds that are based on the belief that by scaffolding guided inquiry classroom experiences and the alignment of standards, meaningful instruction and student understanding will be attained. Students will actually understand the concepts that the teacher is trying to address in the intended curriculum. Classroom discussion and student science notebooks are the tools utilized in this alignment process to help students attain deep understanding of the required science content and to develop complex science reasoning abilities. This alignment also addresses the three guiding principles of how students learn science (National Research Council 2005) by embedding and scaffolding the activation of prior knowledge into the prediction, focusing on the connection of lessons and concepts to lead students to an understanding of the big ideas in science, and by developing metacognitive awareness in students through the use of graphic organizers, Venn diagrams, and other visual devices that assist students in making evidence-based explanations of their conceptual understanding of the science content taught in their classrooms.

The planning structure for alignment must be supported with tools for students such as science notebooks and small-group and classroom discussion to fully attain the depth of understanding that students need in order to become proficient learners. The tools of science notebooks and small-group and class discussion and their importance for students are fully presented in Chapter 2.

2 Tools for Making Meaning
Science Notebooks and Class Discussion

ommunication has a central role in inquiry: it is vital to the progress of science. Words and language are a way of trying out and understanding something (Harlen 2001). When literacy skills are linked to science content, students have a personal and practical motivation to master language and use it to help them answer their questions about the world around them (Thier 2002). Language becomes the primary avenue for arriving at scientific understanding.

The key to effective science teaching is to enable students to develop ideas about the world around them that fit evidence they have collected and imbued with personal meaning. Learning science involves the processes of thought and action and the ability to communicate those thoughts and actions. This learning not only contributes to children's ability to make better sense of things around them but also prepares them to deal more effectively with wider decision making and problem solving in their lives. When students apply the scientific process using the tools of reading, writing, listening, and speaking, they develop the ability to examine their own scientific learning and thinking (Glynn and Muth 1994).

Science Notebooks

Effective science instruction gives students a personal opportunity to construct meaning (Klentschy and Molina-De La Torre 2004). An excellent way for them to do this is by making entries in a science notebook about their science experiences, their social interactions related to these experiences, and their reflections on these experiences and interactions. Student science notebooks, thus, may be viewed as transforming knowledge rather than simply relating it.

Science notebooks are the best record of the science taught in classrooms and learned by students and are an excellent ongoing assessment and feedback tool for teachers (Ruiz-Primo et al. 2002). In today's climate of standards, assessment, and accountability, asking students to voice their thinking and the meaning they have made in relation to science instruction increases their achievement not only in science but also in reading and writing (Amaral et al. 2002; Klentschy and Molina-De La Torre 2004).

The act of writing by its nature—it demands organization—enhances student thinking. Science notebooks used well give stability and permanence to student work, along with purpose and form. They are a record of personally valued information. They also help students link new information with prior knowledge (Rivard 1994) and reflect on science content, thereby developing deeper understanding.

The Science–Literacy Connection

Songer and Ho (2005) have identified three challenging aspects of scientific literacy: (1) formulating scientific conjectures based on observed phenomena; (2) analyzing various types of scientific data (evidence); and (3) formulating conclusions based on relevant evidence. Science notebooks are an opportunity for students to learn to meet all three of these challenges in the context of individual lessons and groups of lessons. In their notebooks, students describe their ways of seeing and thinking about science phenomena and constructing and reconstructing meaning, and do so in their own voice. Asking students to write about their own scientific understanding is an important part of teaching for understanding. By expressing their understanding in their own words rather than parroting information from a book or a lecture, students retain information for a longer time and are able to make connections between individual experiences and experiments.

This kind of self-expression takes time. Young children may at first merely draw what they think is going on, perhaps producing cartoon-like images of smiling insects, but this is an important beginning. With time, practice, and opportunity, they will not only observe more closely but also record and illustrate those observations more accurately and in greater detail. The goal is for students to shed their preconceptions and record and illustrate what they have really observed.

Vygotsky (1978) refers to drawing as graphic speech and notes that young students' representations often reflect what they know about the object instead of what they see. Nevertheless, these drawings reveal students' understanding. Students incorporate particular details in an attempt to make a science experience make sense. These drawings are an important first step in developing the science–literacy connection, especially for English language learners.

Although science notebooks have the potential to help students garner real understanding from their classroom science instruction (Aschbacher and Alonzo 2006), the extent of that potential depends on the guidance and scaffolds they are given. Embedding writing prompts within the inquiry process helps students learn how to create a permanent record of classroom science investigations through questioning, predicting, clarifying, and summarizing. Students need especially strong scaffolds in regard to analyzing data and building explanations from evidence (Songer 2003). Prompts like those listed here can be used at first and gradually removed as students become more experienced.

- Today I [we] want to find out _____ .

- I [We] think _____ will happen because _____ .

- I [We] noticed (or observed) _____ .

Scaffolding Science Inquiry Through Lesson Design

- I [We] wonder _____ .

- Today I [we] learned _____ .

- Questions I [we] have now are _____ .

Science Notebooks and Inquiry-Based Instruction

Using writing to promote learning is consistent with the belief that a writer is actively reprocessing concepts and central ideas (Scardamalia and Bereiter 1986). Writing is, first, polishing one's thinking for self-edification and, second, communicating those thoughts to others (Howard and Barton 1986). The first goal of writing is to understand. Writing is an instrument for thinking: it enables students to express their current ideas about science in a form they can examine and think about.

The earlier children start to learn to keep records (elapsed time, distance covered, and similar measurements), the better they will be prepared to make this a natural part of their science activities (Harlen 1988). But a science notebook is more than a record of data students collect, facts they learn, and procedures they conduct. It is also a record of students' reflections, questions, speculations, decisions, and conclusions relative to scientific phenomena (Thier 2002). In other words, a science notebook is a central place in which the student uses language, data, and experience to form meaning.

Thier (2002) suggests two guiding questions that students can use to help them internalize what they learn:

What new ideas do I have after today's activity?

How can I use what I have learned in my everyday life?

Writing in which students reflect on their own conceptions and reconcile them with available evidence is an effective classroom strategy for learning science (Fellows 1994).

Students' ability to contextualize science phenomena and activity in writing depends on their familiarity with the phenomena and equipment and on how long and how often they have been exposed to the material (Amaral et al. 2002). Unfamiliar situations produce statements about what has been observed; students' first entries in science notebooks are usually narratives or recounting of procedures (Amaral et al. 2002). In familiar situations, children write about their experiences with the phenomena and place the investigation in a real-world context (Shepardson 1997).

One caveat about using science notebooks is that they don't necessarily help students find or identify problems (Reddy et al. 1998). Although students may be quite interested in and excited about carrying out science activities, they may not be willing to spend time interpreting the results. Students' science notebooks are stories that unfold as the observed phenomena change, stories molded to fit the children's way of seeing; therefore the stories may not be that of a scientist or a teacher (Shepardson 1997). Teachers need to guide students to be more reflective about their work. Science notebooks have the potential to move students beyond simply completing the task to making sense of the task and to developing scientific thinking.

Transforming Knowledge Through Student Voice

The way that writing is employed and evaluated in the classroom is critical in determining students' perceptions of its potential for helping them learn (Rivard 1994). In their science notebooks, students can describe their ways of seeing and thinking about scientific phenomena, as well as construct and reconstruct meaning, in their own voice, through their own lens of experience (Shepardson 1997).

Much science writing in the classroom is mechanical and passive, directed toward communicating what the student knows to the teacher as an informed audience: filling in the blanks, producing short responses to teacher-generated questions, and recording observations

Scaffolding Science Inquiry Through Lesson Design

and information (Applebee 1984). In the elementary science curriculum, science notebooks may become logs in which children simply record experiments and list results. No meaning is constructed; knowledge is simply told.

Shepardson and Britsch (2001) insist that students' entries in a science notebook must do more than report teacher-expected results. They suggest that students need to frame scientific investigations and phenomena within a familiar internal context related to an imaginary, experienced, or investigative world. Students must use their own combination of writing, drawing, and genre to construct and represent their understanding. This flexibility lets students use science notebooks in ways that are both socially and cognitively appropriate to their developing understanding of scientific phenomena.

Class Discussion

Talk or discussion also helps students make meaning of their ideas. Just having students perform science investigations and keep science notebooks does not necessarily mean they will learn scientific principles and be able to apply scientific reasoning. Talking in small groups is an important way for students to share their ideas about a science phenomenon, formulate questions and predictions, plan and conduct an investigation, and collect data or record observations. It is a social way for individuals to develop and test ideas.

Teachers should also conduct whole-class discussions before and after science investigations. The discussion before the investigation lets students hear (and see, if notes are jotted on a flip chart or chalkboard) the questions and predictions other groups have formulated and the plan of action they have created for their investigation. In the discussion after the investigation (a what-did-it-mean conference), students share their data and observations, identify any patterns they see, and begin to make claims based on the body of data and evidence collected by individuals.

Summary

Teaching science by having children perform science investigations in the context of writing and class discussion is critical to learning science. Science notebooks and classroom discussion promote literacy while allowing students to clarify their emerging theories about science content.

Scaffolding Science Inquiry Through Lesson Design

Examining Intended Curriculum

3

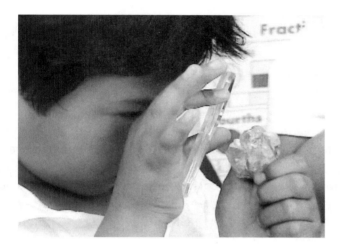

In our daily lives, we usually begin a task with a goal or purpose in mind. It helps us stay on track and monitor our progress. If we apply this principle to the classroom, we see that knowing what to teach is sometimes more important than knowing how to teach. Classroom science teachers in most states are required to follow a predetermined course of study that includes units of instruction, standards, and expected student outcomes. This requires a clear understanding of the content goals for each lesson of every unit.

However, as teachers feeling pressed for time, we often jump into a unit of study by following the teacher's guide without first obtaining a clear understanding of the big ideas or specific content goals. We may also skip lessons we feel are not important or teach lessons out of sequence because we do not understand how one lesson builds on another.

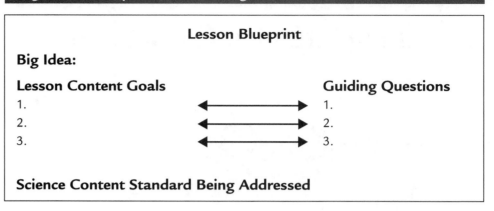

Figure 3–1   Template for Articulating Intended Curriculum

Lesson Blueprint

Big Idea:

Lesson Content Goals   **Guiding Questions**

1.   1.

2.   2.

3.   3.

Science Content Standard Being Addressed

This undercuts the value and effectiveness of our instruction. Students' academic achievement and understanding are enhanced when we identify the specific types of knowledge on which a unit or lesson focuses (Marzano 2003). Stating the explicit focus of each lesson in teacher- and student-friendly terms allows us to concentrate our effort and energy on the essential concept we are asking students to learn. This focused activity helps students be more productive and come to a deeper understanding of what we are trying to teach.

Understanding the *intended curriculum* is the starting point for all lesson planning. It defines the direction of our instruction and the expected outcomes relative to student understanding. The careful crafting of what is intended to be taught lays the groundwork for what is actually taught (the *implemented curriculum,* discussed in Chapter 4).

The template shown in Figure 3–1 can help you articulate your intended curriculum. It has four parts: (1) the big idea, (2) lesson content goals, (3) guiding questions related to these content goals, and (4) the national, state, or district standard(s) being addressed.

Part 1: The Big Idea

Science teachers talk a lot about big ideas and getting students to understand them, but just what *is* a big idea? Essentially, it is an open-ended underlying principle relating to the scientific phenomenon being examined. It unifies and connects the cumulative knowledge and content provided in a series of lessons, all of which share the same big idea. For example, in a unit on magnetism and electricity, the big idea might be that magnetism and electricity are aspects of a single force.

Over time, curricular materials may be revised, but big ideas endure and have lasting value. Our goal as teachers is that after the facts are forgotten or changed, the big idea remains a part of students' knowledge. Therefore, the big ideas need to be public throughout the unit—posted in the classroom on chart paper or written on the board. Also, at the end of each lesson, we need to connect the newly learned concepts and content to the big idea. Doing so enables students to make meaning.

Knowledge of the big ideas in science is the starting point for understanding the flow of the unit, the building blocks established in each lesson, and the overall outcome for students. Here are some examples of big ideas taken or adapted from the National Science Education Standards (National Research Council 1996):

Earth Science Big Ideas

1. Rocks and minerals have different properties that are determined by how they are formed and that make them useful in different ways.

2. Energy from the Sun heats the Earth unevenly, causing air movements that result in changing weather patterns.

3. Water on Earth moves between the oceans and land as a result of evaporation and condensation.

Physical Science Big Ideas

1. Electrical energy can be converted into heat, light, sound, motion, and a magnetic field.

2. As a form of energy, light has a source and travels in a direction.

3. Substances have characteristic properties. A mixture of substances can often be separated into the original substances based on the properties of the substances. Substances react chemically with other substances in characteristic ways to form new substances.

Life Science Big Ideas

1. Living systems at all organizational levels—cells, organs, tissues, organ systems, whole organisms, and ecosystems—demonstrate the complementary nature of structure and function.

2. Organisms have basic needs and can survive only in environments in which their needs can be met.

3. Plants and animals have life cycles that are typical of their species.

Part 2: Lesson Content Goals

Lesson content goals denote the specific outcomes for each lesson in a unit—the science students will learn. They are the building blocks of scientific knowledge. When several lessons are combined, these outcomes form subconcepts. When subconcepts are combined, they form big ideas. When content goals are merely implied within the lessons or located on a remote page in the teacher's guide, their relevance can easily be overlooked.

Part 3: Guiding Questions

Once we have clearly identified the lesson content goals, we can determine what students should know or be able to do by the end of this lesson and express this understanding as related guiding questions. The corresponding content goal is the answer to the guiding question. Guiding questions focus instruction, guide the development of claims and evidence (what students have learned from the investigation, along with evidence that supports this learning), and provide a logical means of assessment.

Below are several examples of lesson content goals and aligned guiding questions that help develop an understanding of a big idea.

Scaffolding Science Inquiry Through Lesson Design

Earth Science Lesson

Big Idea: Rocks and minerals have different properties that are determined by how they are formed and that make them useful in different ways.

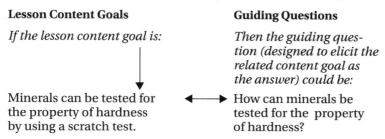

Lesson Content Goals

If the lesson content goal is:

Minerals can be tested for the property of hardness by using a scratch test.

Guiding Questions

Then the guiding question (designed to elicit the related content goal as the answer) could be:

How can minerals be tested for the property of hardness?

Physical Science Lesson

Big Idea: Substances have characteristic properties. A mixture of substances can often be separated into original substances based on the properties of the substances. Substances react chemically with other substances in characteristic ways to form new substances.

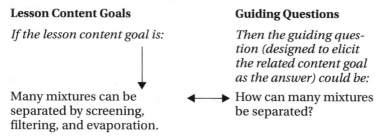

Lesson Content Goals

If the lesson content goal is:

Many mixtures can be separated by screening, filtering, and evaporation.

Guiding Questions

Then the guiding question (designed to elicit the related content goal as the answer) could be:

How can many mixtures be separated?

Physical Science Lesson

Big Idea: Magnetism and electricity are aspects of a single force.

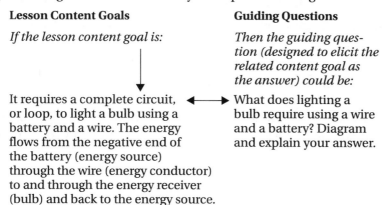

Lesson Content Goals

If the lesson content goal is:

It requires a complete circuit, or loop, to light a bulb using a battery and a wire. The energy flows from the negative end of the battery (energy source) through the wire (energy conductor) to and through the energy receiver (bulb) and back to the energy source.

Guiding Questions

Then the guiding question (designed to elicit the related content goal as the answer) could be:

What does lighting a bulb require using a wire and a battery? Diagram and explain your answer.

Part 4: Science Content Standard(s) Being Addressed

Classroom teachers are also required to cover specific science content standards, as established by their state or school district, during specific grade levels or spans. Often these content standards are the basis for what is included on state tests. Identifying the science content standard(s) being addressed in each lesson helps ensure that the specific content required to be taught is actually taught. Science content standards are often expressed differently from the way lesson content goals are stated and are sometimes broader.

Student understanding of a standard often develops sequentially over a span of several lessons. For example, a state standard may require students to understand that the properties of minerals include hardness, luster, and streak; the specific content goal for an individual lesson may deal only with hardness, with subsequent lessons addressing the other properties. Or students may be required to know that differences in chemical and physical properties of substances are used to separate mixtures and identify compounds; the specific content goals for a series of individual lessons may focus only on physical properties, with subsequent lessons addressing chemical properties.

Putting the Parts Together

The four parts of the lesson blueprint fit together like pieces of a puzzle, forming a road map to help us clearly understand the direction of a lesson. Where is this lesson supposed to go? What is it intended to accomplish with respect to student understanding? Following are a number of examples you can use as models as you plan your own instruction using the Lesson Blueprint Planning Template provided in Figure 3–2.

Scaffolding Science Inquiry Through Lesson Design

Figure 3–2   Lesson Blueprint Planning Template

Lesson Blueprint

Big Idea:

Lesson Content Goals		**Guiding Questions**
1.	←———————→	1.
2.	←———————→	2.
3.	←———————→	3.

Science Content Standard Being Addressed

Lesson Blueprint

Big Idea: Substances have characteristic properties. A mixture of substances can often be separated into the original substances based on the properties of the substances. Substances react chemically with other substances in characteristic ways to form new substances.

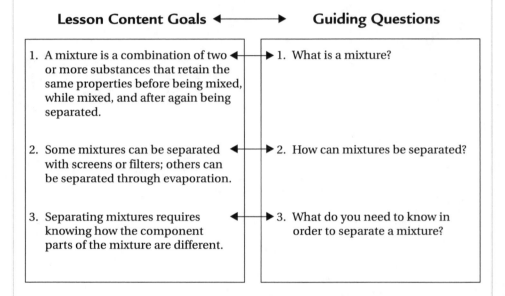

Lesson Content Goals ◄────────► Guiding Questions

Lesson Content Goals	Guiding Questions
1. A mixture is a combination of two or more substances that retain the same properties before being mixed, while mixed, and after again being separated.	1. What is a mixture?
2. Some mixtures can be separated with screens or filters; others can be separated through evaporation.	2. How can mixtures be separated?
3. Separating mixtures requires knowing how the component parts of the mixture are different.	3. What do you need to know in order to separate a mixture?

Science Content Standard Being Addressed

Students understand that differences in physical properties of substances can be used to separate mixtures.

Life Science Intended Curriculum: Decomposers

Lesson Blueprint

Big Idea: All organisms need energy and matter to live and grow. Living organisms depend on one another and their environment for survival.

Lesson Content Goals ◄———► Guiding Questions

Lesson Content Goals	Guiding Questions
1. On land, decomposers such as mold and earthworms break down waste materials, returning them as nutrients to the soil for plants to use.	1. How do decomposers such as mold and earthworms provide nutrients to plants?
2. In water, microorganisms eat waste material and dead animals, and are then eaten by larger organisms, providing them with nutrients and food.	2. How do microorganisms that live in water provide nutrients for larger organisms?
3. Bacteria and protists are beneficial to us because they are the primary producers of biomass in the oceans.	3. How are microorganisms beneficial to us?

Science Content Standards Being Addressed

- Students know that decomposers—including many fungi, insects, and microorganisms—recycle matter from dead plants and animals.
- Students know that most microorganisms do not cause disease and that many are beneficial.

Earth Science Intended Curriculum: Landforms

Lesson Blueprint

Big Idea: Waves, wind, water, and ice shape and reshape the Earth's land surface.

Lesson Content Goals ←——————→ **Guiding Questions**

Lesson Content Goals	Guiding Questions
1. The wearing-away action of water, wind, or glacial ice (erosion) causes slow changes in the Earth's surface.	1. What processes cause the Earth's surface to change slowly?
2. Landslides, volcanic eruptions, and earthquakes cause rapid changes to the Earth's surface.	2. What processes cause the Earth's surface to change rapidly?
3. Erosion caused by flowing water reshapes the land by transporting pebbles, sand, and silt and depositing them in other places.	3. How does land get shaped by flowing water?
4. Besides the weather, rocks are broken down by natural forces such as freezing and thawing and root expansion.	4. Besides the weather, what other natural forces are responsible for breaking rocks into smaller pieces?

Science Contents Standards Being Addressed

- Some changes in the Earth's surface are caused by slow processes, such as erosion; other changes are caused by rapid processes, such as landslides, volcanic eruptions, and earthquakes.

- Natural processes, including weather conditions and root expansion, cause rocks to break into smaller pieces.

- Moving water erodes landforms, reshaping the land by taking pebbles, sand, silt, and mud away from some places and depositing them in other places (weathering, transporting, and depositing).

Setting the Stage for Learning

4

Establishing an intended curriculum is the necessary first step, but that curriculum now needs to be implemented. Students must be given opportunities to acquire the science skills that ultimately lead to understanding science's big ideas.

The implementation process summarized in Figure 4–1 presents instruction in four sequential phases. The first phase sets the stage for learning by developing students' vocabulary through kit inventories and word walls. In the second phase, students formulate investigable questions and make predictions. In the third phase, students plan and conduct their investigations. In the final wrap-up and assessment phase, students provide evidence-based explanations of what they have learned and summarize the meaning they have made from their investigations.

Science notebooks are a key component of this process. As the students move through the implementation phases, they use these notebooks to record the investigable questions and predictions they

Figure 4–1   Sequential Tasks in Implementing Curriculum

Phase 1: Set the Stage for Learning
Vocabulary Development
- ❖ kit inventory
- ❖ dynamic word wall

> **Discussed in this chapter**

Phase 2: Formulate Investigable Questions and Predictions
- ❖ engaging scenario
- ❖ focus question (notebook entry)
- ❖ prediction (notebook entry)

> **Discussed in Chapter 5**

Phase 3: Plan and Conduct the Investigation
- ❖ sequential plan (notebook entry)
- ❖ data collection methods and forms (notebook entry)
- ❖ data obtained (notebook entry)

> **Discussed in Chapter 6**

Phase 4: Make Meaning
- ❖ discussion/analysis (conference)
- ❖ claims and evidence (notebook entry)
- ❖ conclusions (notebook entry)
- ❖ reflection (notebook entry)

> **Discussed in Chapter 7**

have formulated with regard to an investigation, state their plan for conducting their investigation, enter the data they generate during the investigation, make claims based on the investigation and provide supporting evidence, state their conclusions, and reflect on what they have learned.

A typical investigation includes the following elements (the ones in bold become student science notebook entries):

1. Developing an engaging scenario

2. Articulating a **focus question**

3. Making a **prediction**

4. Planning and organizing the **materials and procedures** of the investigation

5. Developing appropriate **data collection/representation forms**

6. Conducting the investigation and recording **observations**

7. Holding a "making meaning conference"

8. Sharing **claims and evidence** statements

9. Revisiting the big idea in relation to the **conclusions**

10. Looking back at the investigation and the learning it precipitated in a **reflection**.

A typical error teachers make is to assume that "all good things must be in every lesson." Remember, an individual lesson won't contain all of these elements, and they needn't be encountered in this strict sequence. Teachers often have students work like scientists, moving back and forth among processes. For example, some investigations into plant growth and development may begin with a prediction of what students think will happen, focus on observing the plant's appearance and measuring its growth, and conclude with a confirmation or revision of the prediction. Which elements to include in which lessons is a matter of professional discretion. The key is to make an informed decision. Will inclusion of the component support student learning? Will its exclusion diminish students' opportunities to make meaning of the intended curriculum?

Working in the Underpinnings

When implementing science curriculum, teachers must (1) help their students develop a positive attitude toward science; (2) teach the skills inherent in the scientific process; (3) employ research-based teaching strategies; and (4) make multidisciplinary connections. Although these elements are discussed separately here, they are interdependent and inseparable from content and big ideas. Process skills are interwoven with content, and students' attitudes toward science are greatly influenced by how process skills are used in active learning.

Student Attitude

Effective instruction includes developing the following student attitudes (Harlen 2001):

- Curiosity (questioning, wanting to know)

- Respect for evidence (open-mindedness, willingness to consider conflicting evidence, perseverance)

- Flexibility (recognition that ideas are tentative)

- Critical reflection (in order to refine and deepen one's ideas and improve one's performance)

- Sensitivity to living things and the environment

Each of these attitudes affects student motivation and determines the way students approach science both in the classroom and later as informed citizens in society. Teachers need to recognize instilling these attitudes as an essential goal.

The Scientific Process

The scientific process is how science is conducted; it is what scientists do. The National Research Council (2005) offers this comparison of the work of scientists and the work of students conducting guided inquiry:

Work of Scientists	Work of Students Conducting Guided Inquiry
• Posing a research question • Formulating hypotheses • Designing investigations • Making observations, gathering and analyzing data • Building or revising scientific models • Evaluating, testing, or verifying models	• Posing a research question • Formulating predictions • Designing investigations • Making observations, gathering and analyzing data • Building or revising scientific models • Evaluating, testing, or verifying models

Scaffolding Science Inquiry Through Lesson Design

Because the work is identical, student science notebooks should provide evidence of the following activities:

- Questioning
- Predicting
- Planning/organizing investigations
- Collecting data (evidence)
- Making claims supported by evidence (data)
- Drawing conclusions and reflecting
- Communicating results.

Research-Based Teaching Strategies

Research-based teaching strategies are the most effective relative to student success and include an array of choices. The following list is based on an extensive review of the research literature, and the strategies are directly applicable to science notebooks (Marzano et al. 2001).

- Identifying similarities and differences
- Summarizing and taking notes
- Reinforcing effort and providing recognition
- Using nonlinguistic representations (drawings, charts, graphs, illustrations)
- Participating in cooperative learning experiences
- Setting objectives and providing feedback
- Generating and testing hypotheses or predictions
- Using questions, cues, and organizers.

We should consider each of these strategies in relation to the purpose of a lesson, how we want to implement the lesson, and what we want students to know and be able to do as a result of the lesson.

Multidisciplinary Connections

Science may be one of the best content areas in which to make multidisciplinary connections. Scientific investigations in which students read, write, listen, and speak have explicit connections to language arts. Literacy is the foundation for all learning, and language—communication—is the means by which most content is delivered.

Each science lesson provides opportunities to reinforce language arts and math essentials. As students write in their science notebooks, they are supported by word walls, discussion prompts, sentence scaffolds, and organizational aids. Science notebooks provide the data and rich academic language necessary for effective communication, both oral and written. When students collect and analyze quantitative data, they are using mathematical concepts in an authentic application.

Using Kit Inventories and Word Walls

When we recognize that student attitude, the scientific process, and multidisciplinary connections need to be embedded in the curriculum, we become very aware that lessons can't just appear out of nowhere. As a runner warms up before embarking on a 5K run, students need to warm up before instruction begins if they are to receive the maximum benefit of that instruction. Activities such as kit inventories and word walls establish a common vocabulary and introduce the materials that will be used in the unit or lesson.

Kit Inventories

In taking a science kit inventory, students ask questions about and discuss the scientific name, use(s), and properties of each item. This is an excellent opportunity for students to build vocabulary in a coherent, contextual way. Becoming familiar with each item's use or function, its origin, and how it relates to a scientific concept paves the way for students to be able to make meaning from an investigation. Exposing students to the appropriate vocabulary at the beginning of a unit gives them a common language to use when they write in their science notebooks and discuss concepts with others. (This is especially important in language-diverse classrooms.)

Scaffolding Science Inquiry Through Lesson Design

There are various methods for inventorying a science kit. One is to introduce each item individually and ask students to identify it (or relate it to another object with which they are familiar) and then predict how the item might be used in the unit of study. A sample discussion about the items in a kit used in a rocks and minerals unit is shown in Figure 4–2.

A teacher may tap into the students' prior knowledge by asking, "Where have we seen this before?" and "How have you seen it used?" The discussion that develops helps the teacher assess student understanding of relevant vocabulary. Asking students to describe each

Figure 4–2   Sample Kit Inventory Questions from a Unit on Rocks and Minerals

Item	Teacher	Students
Ruler	What is this called?	A ruler.
	What is it made of?	Plastic.
	Where have you seen it before?	At home.
	Why do people use it?	To measure things.
	Why do you think rulers are in our science kits?	To measure things.
Nail	What is this called?	A nail.
	Where might you use one of these?	To fasten two pieces of wood together.
	How do the ends of the nail feel?	One end is pointed and sharp and the other end is smooth and flat.
	How might we use a nail in our science unit?	To use the sharp pointed end to make holes in the rocks.
Glass droppers	What is this item called?	A glass dropper.
	Where have you seen this before?	At home in our bathroom.
	How might we use these in our science unit?	Maybe to measure liquids?

item calls attention to its physical properties. (For a more detailed explanation of kit inventories, see Amaral et al. 2006.)

After discussing each item in the science kit, you might place the necessary quantities of each item in plastic bags and pin them to a bulletin board under a label bearing the name of the item (see Figures 4–3 and 4–4).

Figure 4–3   Science Kit Display Board

Figure 4–4   Another Science Kit Display Board

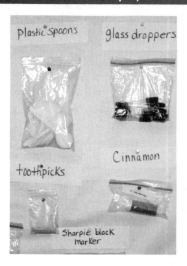

Scaffolding Science Inquiry Through Lesson Design

Dynamic Word Walls

The relationship between vocabulary and academic achievement is well established (Marzano 1991). Word walls display words directly related to items and concepts in the lesson being taught and have a profound effect on students' ability to understand the lesson. Words, definitions, and illustrations are added lesson by lesson as students participate in a direct experience or a structured discussion before making a science notebook entry. Connections between new words and familiar words are indicated with arrows, pictures, and symbols. The goal of creating a word wall is for students to use rich academic language as they listen, speak, read, and write. Word walls are more than a list of words and definitions; that why I've added the word *dynamic*. The entries contain information that connects what students already know with new concepts. (See the examples in Figures 4–5, 4–6, and 4–7.)

Word walls and charts are extremely important for all students, but even more so for English language learners, who thus have a visual reference to use during class discussions, while participating in group work, and when writing in their notebooks. (For more information about word walls, see Duron-Flores and Macias 2006.)

Having laid the foundation for a lesson by introducing vocabulary and discussing relevant tools and materials, we can now move on to Phase 2.

Figure 4–5 Earth Materials Dynamic Word Wall

Words Rock!

Mineral Made of one ingredient; cannot be separated into anything other than what it is no matter how small you make it. Examples: calcite, fluorite, quartz, and gypsum.

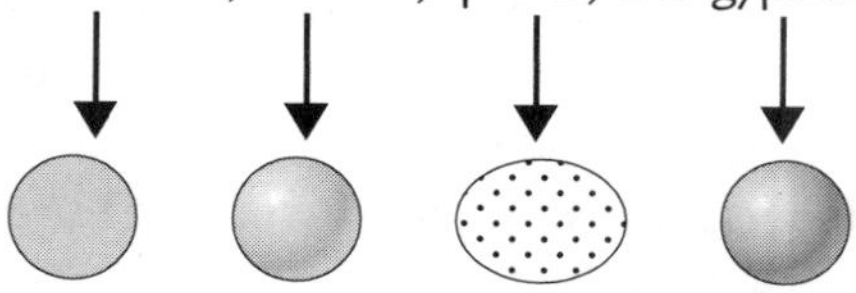

Scratch test Determining the **hardness** of a mineral by the mark left after it is scratched with a tool or another mineral.

Seriate To arrange in series or order.
Example: lightest to heaviest or softest to hardest.

Curator Person in charge of a museum or library.

Properties Characteristics we can see or observe about a mineral, such as shape, size, color, texture, **hardness,** luster.

Tools Used to scratch our minerals: fingernail, penny, and paper clip.

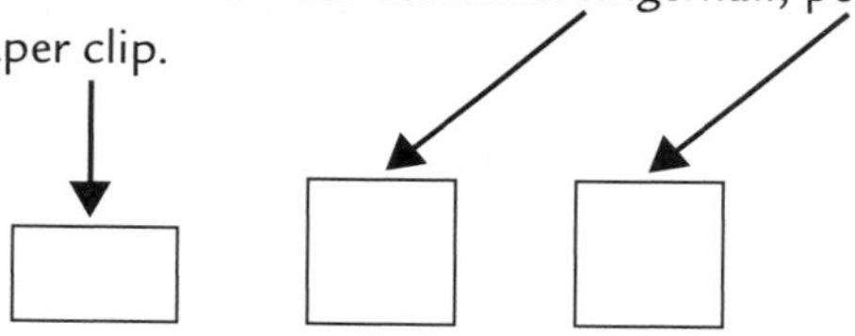

hardest tool   softest tool   medium hard tool

Claim What we know or learned based on our data.

Evidence How we know what we claim—what did we observe or measure?

Hand lens Tool used to get a magnified (close-up, bigger, more detailed) view.

Figure 4-6 Electricity and Magnetism Dynamic Word Wall

Watt's My Word?

Battery

negative   positive

The energy source
- • = Critical contact point—very important place to make contact/touch

Bulb Energy user—is evidence of a complete or open circuit if it lights or does not light.
What are examples of energy users?

Wire Completes the pathway between the energy source and the energy user—wires are metal (copper).

Complete circuit A complete/closed loop from one end of the power source (battery) through the energy user (bulb) and back to the other end of the battery; evidence of a complete circuit is a lit bulb.

> Labeled diagram of a complete circuit with the circuit traced in red with arrows starting at the negative end of the battery.

Open circuit A circuit that does not make a complete loop—in an open circuit the bulb does not light.

> Labeled diagram of an open circuit—trace the path in red only as far as it goes.

Conductor Material that allows the flow of electricity.
Examples: metals such as copper.

Insulator Material that does not allow the flow of electricity.
Examples: glass, plastic, wood.
Why are insulators important?

Inside a Bulb

Support wire

• = Critical contact point

Filament—thin wire
Glass bead
Support wire
Side terminal
Base terminal

What is the role of the filament, glass bead, and support wires?

Figure 4–7   Mixtures Dynamic Word Wall

All Mixed Up? Words and More Words!

Mixtures
Two or more materials mixed evenly together; they retain (keep) their original properties after being separated.

> Examples of mixtures:
> Keep an ongoing list—have students add examples of mixtures that have not been mentioned in class as a way of determining their conceptual understanding

Gravel
| sample |

Small bits of hard rock.
What are some uses for gravel?

Salt
| sample and
a diagram of a salt
crystal |

Mineral in form of a crystal (a solid with the special shape of a cube).
How is salt used? Where is it found?

Diatomaceous earth (powder)
| sample |

Hard, white, gritty shells of microscopic organisms called diatoms.
What are some uses for diatomaceous earth?

Properties
Observable or measurable characteristics.

Screen
| sample |

Open mesh that keeps particles larger than the mesh openings on the top of the screen and allows particles smaller than the openings to fall through.

Filter paper
| sample |

A screen with very small openings/holes.

How are a screen and filter paper the same? Different?

Formulating Investigable Questions and Predictions

5

The second phase in implementing curriculum (see Figure 5–1) is to help students formulate investigable questions and predictions.

Engaging Scenarios

Student learning is triggered by stories or narratives designed to arouse students' curiosity and prompt them to develop investigable questions and predict answers. Situations common to everyday experience set the context for solving problems in scientific ways by posing a challenge that becomes the focus of a lesson. Figure 5–2 is an engaging

Figure 5–1   Sequential Tasks in Implementing Curriculum

Phase 1: Set the Stage for Learning
Vocabulary Development

❖ kit inventory

❖ dynamic word wall

> Discussed in
> Chapter 4

Phase 2: Formulate Investigable Questions and Predictions

❖ engaging scenario

❖ focus question (notebook entry)

❖ prediction (notebook entry)

> Discussed in
> this chapter

Phase 3: Plan and Conduct the Investigation

❖ sequential plan (notebook entry)

❖ data collection methods and forms
  (notebook entry)

❖ data obtained (notebook entry)

> Discussed in
> Chapter 6

Phase 4: Make Meaning

❖ discussion/analysis (conference)

❖ claims and evidence (notebook entry)

❖ conclusions (notebook entry)

❖ reflection (notebook entry)

> Discussed in
> Chapter 7

Figure 5–2   Third-Grade Earth Materials Scenario

Yesterday my friend who lives next door told me about her trip to her grandfather's cabin in the mountains. While she was there, she found some interesting rocks, and she wanted me to see them. Carefully, she removed her treasures from a shoebox. She thought maybe some might be valuable—that some might even be minerals. She asked me what I thought.

"Well," I told her, "last year in science we learned that rocks are really collections of minerals. Since some of your rocks have many different colors, they are probably made of different minerals."

But a few rocks were all the same color. My friend reminded me that minerals are usually one color and have distinctive characteristics. We thought we might get some clues from a chart about rocks that I had and be able to identify the rocks and perhaps the minerals in them.

If you were asked to help, could you do it? What is the problem that this investigation might try to solve?

scenario for a third-grade lesson on earth materials. Figure 5–3 is an engaging scenario for a fourth-grade unit on magnetism and electricity. Figure 5–4 is an engaging scenario for a fifth-grade unit on mixtures and solutions.

Figure 5-3   Fourth-Grade Magnetism and Electricity Scenario

You are out for a nature hike with your friends and meet a park ranger at a campsite. He tells you something funny has just happened. While he was packing up some things at the camp, he pushed some plastic bags around on the ground. When he lifted one of the bags, which was empty, he saw several pieces of paper attached to it. Then he noticed a piece of paper stuck to a plastic apron he had on. The pieces of paper clung to the bag and to his apron until he took them off, and when he removed them they made a crackling sound. He thinks maybe the phenomenon is related to magnets and electricity, and he asks you to help him find out why this happened. There are some additional plastic bags on the picnic table you can use to help you.

What problem needs to be investigated here?

Figure 5-4   Fifth-Grade Mixtures and Solutions Scenario

Earlier today you and your friend were in your family's garage planning a science lab for your little brothers and sisters.

"Let's do sink and float," your friend suggested.

"Great idea!" you responded, and the two of you immediately began gathering materials you found on the workbench. Then you filled some jars with water and placed the materials into the jars to see if they would sink or float. Some floated, some sank, and to your surprise, some of the materials actually seemed to disappear into the water.

At dinner (your mom invited your friend to stay and eat with you), your dad starts talking about some materials he has on his workbench that are for a project at school (he is a professor at the university). He says he plans to show them to his students tomorrow. You and friend look at each other nervously. You've ruined your dad's project!

After dinner you and your friend go back to the garage to try to fix what you have done.

What is the problem you must solve?

Focus Question

At the heart of investigative science is a scientist's ability to ask questions. Classroom teachers use engaging scenarios like these examples as motivation—a "hook"—to get students to formulate their own investigable questions. At the conclusion of each engaging scenario, they ask, "What is the challenge to which a solution must be found? What is the question that must be answered or solved?"

Because the development of a high-quality focus question is the key to a good investigation, it is very important that students are given criteria with which to develop those questions (see Figure 5–5).

Formulating investigable questions takes time and practice. The less experience students have in formulating investigable questions, the more scaffolding teachers need to provide. Researchers from the California Institute of Technology have developed a continuum of scaffolding strategies teachers should consider when planning effective science lessons (Aschbacher and McPhee-Baker 2003; see Figure 5–6).

What exactly does an investigable question look like? Most start with *what* or *how*:

What do I want to find out?

What is the reason for my question?

What problem am I addressing?

What would happen if . . . ?

How can we . . . ?

Figure 5–5 Criteria for a High-Quality Focus Question

1. Does the focus question allow for an investigation that directly relates to the engaging scenario and to the intended curriculum of which it is a part?
2. Does the focus question lead to a investigation comprising coherent steps and an opportunity to communicate results?

Scaffolding Science Inquiry Through Lesson Design

Figure 5–6   Scaffolds for Writing High-Quality Focus Questions

- ❖ Teacher provides focus question.
  This is appropriate when:
  - Students do not have experience generating quality focus questions on their own.
  - Students are unfamiliar with lesson content/materials.
  - Content knowledge or some other goal is most important.
- ❖ Teacher provides two (or more) focus questions for the class to discuss and then choose the best one.
- ❖ Teacher facilitates a class discussion during which students' suggested focus questions are critiqued. Then:
  - The class chooses a single focus question.

    or

  - The class chooses several possible focus questions, from which individual students choose one.
- ❖ Students write their own focus questions and display them on a chart, overhead transparency, or whiteboard. The questions are then discussed with and critiqued by the class/a partner/their table group. Students who need more support can use a focus question generated by a classmate.

For example, a fifth-grade class was studying the circulatory system. Because many of the students had little or no experience formulating their own focus questions, the teacher decided to model the process by having them generate a focus question as a class. After a lot of discussion, the class came up with the following question:

Lesson 3 Day 1 11/7/05 1:47 pm
Class focus Question!
Where does blood with oxygen and blood
with carbon dioxide and wastes go in the
body, and where do they come from?

Based on this question, groups of students planned and conducted investigations of the circulatory and respiratory systems and learned the function of the heart, lungs, pulmonary arteries, and veins. The students also learned how oxygen and carbon dioxide in the blood are exchanged via the respiratory system and why we inhale and exhale as part of this process.

Following is a student-generated investigable question for a third-grade lesson on putting minerals in order from hard to soft using a scratch test and the Mohs' scale.

Focus Question

How can we describe the minerals? How can we put them in order from soft to hard?

A second student-generated focus question focused on rock (mineral) formation.

Focus Question:

How can we classify the rocks in the bag according to how they are formed?

The scenario in Figure 5–8 relative to magnetism and electricity posed the problem of how to light a bulb using a piece of wire, a battery, and a bulb. One student suggested the focus question shown next.

Can you use a battery to turn on the lightbulb?

Scaffolding Science Inquiry Through Lesson Design

This question is not an effective focus question, because it can be answered yes or no. The teacher prompted the student to reconsider his question.

Teacher: Is this a question that can be answered yes or no?
Student: Yes.
Teacher: What would your question look like if you used one of the sentence starters on the whiteboard: What would happen if . . . ? How can we . . . ?
Student: How can we use a battery to turn on the lightbulb?
Teacher: I think that is a great question.

Another student suggested the following focus question.

Lesson 1:
Focus Qustion: How can we make & light bulb out of these materials.

The teacher also prompted this student to reconsider.

Teacher: Are you going to make a lightbulb in your investigation?
Student: No, I want to make the bulb light up.
Teacher: What materials are you going to use?
Student: A wire, a bulb, and a battery.
Teacher: So how can you restate your question to describe what you just shared with me?
Student: How can I use a battery, a wire, and a bulb to make light?
Teacher: That's a really good question!

Later in the same unit on magnetism and electricity, a student wrote the following investigable focus question regarding the common properties of a magnet and a compass.

Focus Question
Focus Question: What properties do a compass and a magnet have in commen?

A fifth grader wrote this investigable question regarding making mixtures and then separating them into their component parts.

Focus Question?
How can I make a mixture and then separate it into its original parts?

The goal is for students to write their own focus questions. Questions that begin with *what* or *how* lead students to rich investigations that will best help them attain the content goals identified in the intended curriculum.

Students should enter their focus question in their science notebook after they have discussed it with a partner or a small group. They are then prepared to share the focus question during whole-class discussion, and the question will become a starting point for the rest of the lesson.

Prediction (Notebook Entry)

Students should also predict what the answer to the focus question will be. These predictions give teachers insight about students' prior knowledge and misconceptions and give students a stake in the results. A class of third graders was asked to predict the ways in which minerals could be classified. One group of students predicted that size was the best way to classify minerals, because one could easily place the minerals in order from largest to smallest. These students held the misconception that the minerals they were working with were always the same size and not just pieces of a larger form of the mineral. In a fourth-grade class, students predicted that just the raised "nipple" on the positive side of a battery, not the entire surface, was the point of contact.

Prediction strategies in science are closely linked to prediction strategies in reading in that both situations require students to determine what they think will happen next (Klentschy and Molina-De La Torre 2004). In reading, students make these predictions based on their prior experiences in reading stories of the same genre, their life experiences, and what they would *like* to have happen in the story. In science, classroom teachers should encourage students to make predictions about what they think will happen in the investigation based on the problem posed in the engaging scenario, their focus question, and their prior experiences.

We don't want students to make "wild guesses" that are unrelated to the problem posed in the engaging scenario or to their focus question. Predicting is strengthened through various kinds of scaffolding (Hug et al. 2005), which include the following formulations:

I think or predict that ——————— , because ——————— .

If ——————— , then ——————— , because ——————— .

Requiring students to use the word *because* prompts them to explain how they think their investigation will solve the problem posed in the engaging scenario. It also activates their prior knowledge and ensures a more reasonable prediction (National Research Council 2005).

Following is the prediction the student produced for his mineral focus question presented earlier.

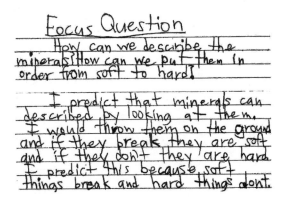

The *because* clause signals an attempt to activate prior knowledge, even though the reason stated reveals a lack of experience in testing the hardness of minerals and rocks.

The earlier investigable focus question related to making a complete circuit with a wire, battery, and bulb generated the next prediction.

Focus Question

How do we make light with a wire, a battery, and a lightbulb?

Prediction

If you attach one end of the wire to the bottom of the bulb and the the other end of the wire to the bottom of the battery then the bulb will light. I predict this because I know a battery can light a bulb.

This student also uses the word *because* and demonstrates a basic (although incomplete) understanding of a circuit and the role of the battery, wire, and bulb in completing it. It is hoped that through the investigation, this student will come to understand the importance of contact points and the similarity of a circuit to a complete loop or circle.

Here's the prediction related to the focus question regarding the similarities and differences between cheek and onion cells.

✗ What are the similarity and differences of the cheek and onion cells?

I think that if I use the microscope I expect to see many differences because the onion cell comes from a vegetable and the cheek cell is from a human

Again, the word *because* activates the student's prior knowledge of differences between plant and animal cells. Later in the investigation, the student will learn that there are distinct differences in plant and animal cells.

A fifth grader studying the circulatory system used the class focus question for the investigation and then wrote a prediction using the If . . . , then . . . , because . . . prompt.

Scaffolding Science Inquiry Through Lesson Design

> Lesson 3 Day 1 11/7/05 1:47 pm
> Class Focus Question!
> Where does blood with oxygen and blood
> with carbondioxide and wastes go in the
> body, and where do they come from?
>
> Prediction: If veins transport blood with
> carbon dioxide, then I think they are connected
> in some way to the lungs because we breath
> out carbon dioxide.

This student's prior knowledge of the circulatory and respiratory systems included some understanding of the function of veins and the exchange of oxygen and carbon dioxide that takes place when we breathe.

Sometimes teachers have students draw a diagram of their prediction. This visualization strategy is particularly useful for English language learners and other students who may have linguistic challenges. The two diagrams that follow are by fourth graders conducting an investigation on how to make a complete circuit using a wire, battery, and bulb. Each diagram shows an incomplete prior understanding of what makes a circuit but does demonstrate an intuitive understanding of the role of the wire, battery, and bulb.

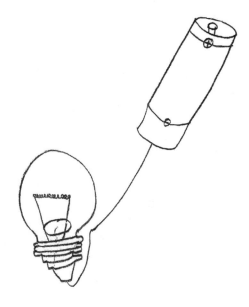

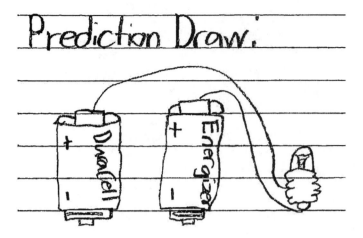

The engaging scenario, focus question, and prediction all play an important role in motivating students and getting them to focus on the lesson content goals identified in the lesson blueprint, and students' focus questions and predictions are significantly strengthened through scaffolding and class discussion. Teachers need to provide opportunities for students to form and discuss focus questions in small groups, write focus questions in their science notebook, and then share and discuss their focus questions in whole-class discussion, applying the criteria for high-quality focus questions. The same process can be applied to predictions.

Figures 5–7, 5–8, and 5–9 are examples of engaging scenarios, focus questions, and predictions related to specific intended curriculum to illustrate how these three processes are related and how they flow seamlessly within the lesson blueprints.

In Chapter 3, you were asked to develop a blueprint—big idea, lesson content goals, and guiding questions—for a lesson you might teach with your own students. Now, taking into consideration what your students already know and do not yet know, write an engaging scenario, focus question scaffolds, and prediction scaffolds for that lesson. Also think about kit inventory strategies to use to introduce the unit, key vocabulary that will be used in the lesson, and how to help students create and use a dynamic word wall. You may use the template in Figure 5–10 to help you.

Scaffolding Science Inquiry Through Lesson Design

Figure 5–7 Earth Materials Lesson (Scratch Test)

Engaging Scenario

You and your group have been hired by the local natural history museum to assemble a display about minerals. The curator (the person in charge) of the exhibit has given you four mineral samples, numbered 1, 2, 3, and 4. She wants the minerals described, compared, and named in the display. Additionally, the minerals need to be seriated (put in order) from softest to hardest. The only accompanying information is a list identifying the four minerals as calcite, quartz, gypsum, and fluorite; however, there is no indication which of the samples provided is which.

Focus Question (Notebook Entry)

If your students are inexperienced at identifying quality focus questions, you can suggest these two possible focus questions.

1. How do we describe, name, and put the four minerals in order of hardness?
2. Can we describe the minerals?

After discussion, question 2 will be eliminated, because it can be answered yes or no. It is also incomplete, because it addresses only part of the required task: the ordering of minerals by hardness is left out.

The students enter question 1 in their science notebook.

Prediction (Notebook Entry)

Prompt: How might you answer the focus question based on what you already know?

Possible Frame

I think/predict that _____

because _____.

Figure 5–8   Magnetism and Electricity Lesson: Lighting the Bulb

Engaging Scenario

While out on a nature hike, you and your group tumble into a dark cave. No one is injured, but it is extremely dark in the cave. Without more light it will be difficult to find your way out. You cannot find a flashlight in any of your backpacks, but you do find a bulb, a battery, and a wire. Will they be of any use to you in creating some light?

Focus Question (Notebook Entry)

Possible Focus Questions

1. What is inside a battery?
2. How can we connect a battery, bulb, and a wire together so the bulb lights?
3. How can we get a bulb to light using only one wire and one battery?

Discussion

Question 1, finding out what is inside the battery, is interesting but not specific to the lesson's intended curriculum. However, this question can be listed on an Interesting Questions chart and explored as an extra-credit or homework assignment.

Questions 2 and 3 both meet the criteria for a quality focus question.

Students enter either question 2 or question 3 in their science notebook.

Prediction (Notebook Entry)

Prompt: How might you answer the focus question based on what you already know?

Possible Frames

I think/predict that _____

because _____.

If _____ , then _____ ,

because _____.

Scaffolding Science Inquiry Through Lesson Design

Figure 5–9   Mixtures and Solutions Lesson: Separating Mixtures

Engaging Scenario

While playing in your family's garage, you and some friends found some gravel, some salt, and a gritty white powder, and you mixed each of them together with water. It was a lot of fun until your father came home and got upset because the salt (halite), gravel, and gritty white powder (diatomaceous earth) were materials he needed for his work. He explained that diatomaceous earth is made up of the hard white shells of microscopic organisms called *diatoms*. None of you have enough money to buy new materials. Now what?

Focus Question (Notebook Entry)

Have students, in cooperative groups, brainstorm possible focus questions and critique them based on the focus-question criteria. Ask each group to select their best question for whole-class review. Generate a class list of questions that meet the criteria. Students may enter any of the selected questions in their science notebook.

(Look for focus questions similar to this one: How can we separate the salt from the water, the gravel from the water, and the diatomaceous earth from the water?)

Prediction (Notebook Entry)

Prompt: How might you answer the focus question based on what you already know?
Possible Frames

I think/predict that _____

because _____ .

If _____ ,

then _____ ,

because _____ .

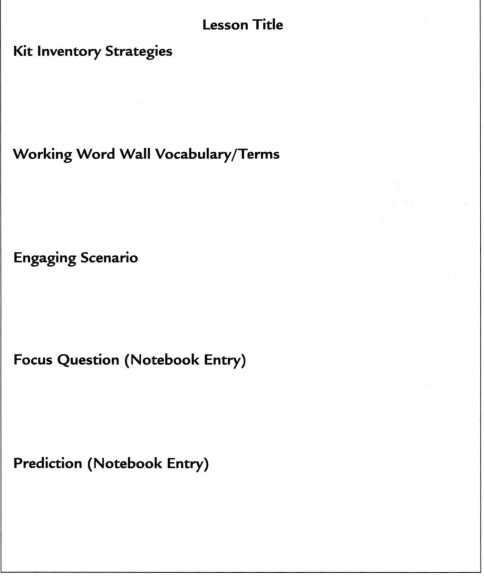

Lesson Title

Kit Inventory Strategies

Working Word Wall Vocabulary/Terms

Engaging Scenario

Focus Question (Notebook Entry)

Prediction (Notebook Entry)

Planning, Organizing, and Conducting the Investigation

6

In the third phase of implementing curriculum (see Figure 6–1), classroom teachers help students plan and organize the investigation and then conduct it. Students collect data and record their observations, then later use this information to formulate claims (evidence-based explanations) and draw conclusions that will enable them to answer the focus question that began it all.

Planning and organizing an investigation is a vital part of guided inquiry, and scaffolding strategies are a necessary component of the process. There are three important steps in planning and conducting an investigation: (1) designing the general plan; (2) designing the operational plan; and (3) developing graphic organizers to aid data collection.

Figure 6–1 Sequential Tasks in Implementing Curriculum

Phase 1: Set the Stage for Learning
Vocabulary Development

❖ kit inventory

❖ dynamic word wall

| Discussed in Chapter 4 |

Phase 2: Formulate Investigable Questions and Predictions

❖ engaging scenario

❖ focus question (notebook entry)

❖ prediction (notebook entry)

| Discussed in Chapter 5 |

Phase 3: Plan and Conduct the Investigation

❖ sequential plan (notebook entry)

❖ data collection methods and forms (notebook entry)

❖ data obtained (notebook entry)

| Discussed in this chapter |

Phase 4: Make Meaning

❖ discussion/analysis (conference)

❖ claims and evidence (notebook entry)

❖ conclusions (notebook entry)

❖ reflection (notebook entry)

| Discussed in Chapter 7 |

Designing the General Plan

Exposing students to a series of questions scaffolds or frames the process of planning an investigation. These questions help students transform their focus questions and predictions into a plan of action leading to the actual investigation. The questions are simple, and with repeated exposure, students will use them automatically:

• What should be changed in the investigation? (This is the independent variable.)

Scaffolding Science Inquiry Through Lesson Design

- What should be kept the same? (These are the controlled variables.)

- What should be observed or measured? (This is the dependent variable.)

As an example of this approach, let's visit a fourth-grade classroom in which students are studying magnetism and electricity. The students have just explored the properties of a simple circuit and are beginning to investigate parallel circuits. The students in one group have written the following focus question and prediction.

Focus Question

How can two lightbulbs glow equally bright with one battery?

Prediction

We think that if the circuit has more than one path, then the lightbulbs will glow equally bright, because it will be able to provide them with the same energy.

Their teacher now provides the group with the following scaffold they can use to help them develop a general plan.

Planning Step	General Plan
1. What should be changed?	
2. What should be kept the same?	
3. How will differences be observed or measured?	

After a group discussion, the students create the following general plan.

Planning Step	General Plan
1. What should be changed?	The number of wires to make two circuits from one battery
2. What should be kept the same?	One battery Two bulbs
3. How will differences be observed or measured?	Are the bulbs the same brightness?

From this general plan, the students are then ready to develop an operational plan for conducting their investigation.

Designing the Operational Plan

Prompted by their teacher's guiding question (What materials and tools will you need?), students list the materials they will use in their investigation, referring as they do to the dynamic word wall they began when the lesson was introduced.

Each working group of students then discusses the procedural questions identified in their general plan. Later, in a general class discussion guided by the teacher, each group shares the gist of these discussion points. The following operation plan results.

Planning Step	General Plan	Operational Plan
1. What should be changed?	1. The number of wires to make two circuits from one battery	1. Use two wires each connected to the positive and negative ends of the battery. Attach each wire to the critical contact points on the bulb.
2. What should be kept the same?	2. The battery and the lightbulbs	2. Use one D cell battery and the same two lightbulbs.
3. How will differences be observed or measured?	3. The brightness of the bulbs	3. Observe whether the two bulbs glow with the same brightness or not.

Some teachers question the effectiveness of having their students enter the operational plan in their science notebooks. Primary teachers in particular feel it takes too much time and may undermine the purpose of the plan. Here's a good rule of thumb: If students are designing their own experiments, the general plan and operational plan need to be documented in the science notebooks. However, if the entire class is conducting the same investigation, then both the general plan and the operational plan can be displayed on a class chart or the dynamic word wall; there is indeed little to be gained from having students copy this information into their science notebook.

If students are designing their own investigation, offering them an effective writing scaffold (*First, . . . second, . . . then, . . . finally, . . .*) will help them make connections to the reading–language arts skill of sequencing.

Making Thinking Visible Using Data Organizers

Having created a plan and decided on the appropriate materials and tools to use in the investigation, students next need to consider how the data will be collected and organized. The third question guiding the design of the general plan—What should be observed or measured?—is an effective starting point for identifying graphic organizers to record the data. Knowing the type of data that will be generated, the students can discuss what types of graphic organizers are appropriate. Typical graphic organizers include charts, concept maps, graphs, T-charts, Venn diagrams, lists, labeled illustrations, and diagrams.

A group of third graders preparing to compare the similarities and differences of two minerals chose to record their data on a Venn diagram (see Figure 6–2). Similarities will be listed in the area common to both circles, differences in the areas unique to each circle.

Figure 6–2   Venn Diagram for Comparing Two Minerals

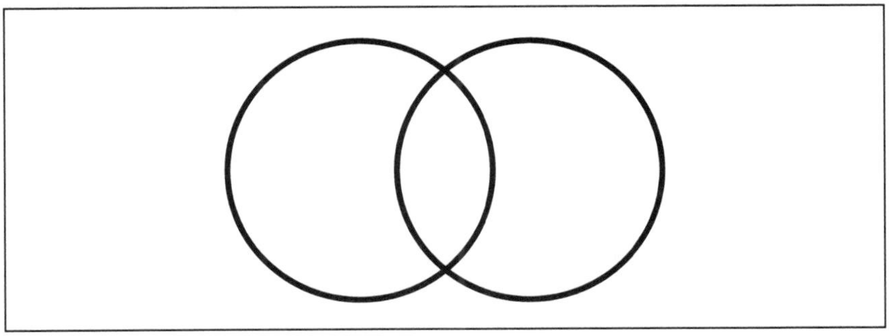

Figure 6–3	T-Chart for Listing Methods That Did and Didn't Work

Ways the Bulb Lit	Ways the Bulb Did Not Light

A fourth-grade group investigating simple circuits chose a T-chart (see Figure 6–3) on which to list the ways the bulb lit and the ways the bulb did not light.

A fifth-grade group developed the chart shown in Figure 6–4 on which to collect data for an investigation on chemical changes.

The patterns and relationships revealed in a graphic organizer help students form conceptual and procedural understanding. Organizers need be modeled and developed with student input. Even if the resulting organizer is virtually identical to one provided by a curriculum developer, having the experience of developing one from scratch is important if students are to become self-sufficient data organizers. Students not only gain a deeper understanding of the purpose of the data collection graphic but also gain experience in

Figure 6–4	Data Observation Chart for Fifth-Grade Chemical Change Investigation

Substances	Physical Properties Observed	I Think the Substance Is . . .
Substances left on filter paper		
Substances left in evaporating dish		

Scaffolding Science Inquiry Through Lesson Design

creating labels for its elements. This helps students better understand and interpret the data that they collect.

Conducting the Investigation and Entering Data

The investigation/exploration itself includes learning tasks aligned with the intended curriculum. Specific roles for the students and the teacher are shown in Figure 6–5. The extended examples in Figures 6–6, 6–7, and 6–8 illustrate scaffolding instruction while planning, organizing, and conducting an investigation as part of an implemented curriculum. After studying these examples, you can use the template in Figure 6–9 to plan, organize, and conduct the lesson/investigation you have been developing in the previous chapters.

Figure 6–5   Student and Teacher Roles During the Investigation

Students' Role	Teacher's Role
❖ Implement the investigation plans.	❖ Use guiding questions from the intended curriculum to focus observations and data collection.
❖ Observe carefully; record observations noting similarities and differences.	❖ Prompt careful observation and encourage note taking:
❖ Work cooperatively.	● What do you notice about _____?
❖ Try alternatives.	● What would happen if _____?
❖ Accurately record data and ideas.	● How is _____ like _____ or different from _____?
❖ Measure and manipulate variables/materials carefully.	● What do you wonder about ____?
	❖ Monitor students as they manipulate variables, measure, and carry out procedures.
	❖ Redirect students when necessary by having them review the focus question.
	❖ Suggest replications.
	❖ Give oral feedback on the process of inquiry and on notebook entries.

Step 1

As requested by the curator of the museum, the students decide to describe the properties of the four minerals.

Procedures

Class discussion prompt: What properties of the four minerals can be described? Answers: color, shape, texture (how it feels when touched).

Materials

All materials are part of the data chart and the word wall.

Data Organization

How can we organize the data?

Notebook Entry (data chart developed with student input)

Museum Display: Observable Properties of Four Minerals

	Color	Shape	Texture	Other
Mineral 1				
Mineral 2				
Mineral 3				
Mineral 4				

Discussion

1. Descriptions are shared and entered on a large class organizer.
2. What properties of minerals can be described by observation?
3. What kind of data/information do we need to seriate (put in order) the hardness of the minerals from softest to hardest? Does the information on this chart help us do this?

Result

Students decide the information on the data chart doesn't enable them to name the minerals or put them in order of hardness. They need more information.

Figure 6-6 *continued*

Step 2

Students decide to ask a geologist (a person who studies the Earth and the Earth's materials) at the museum for information about testing the hardness of minerals. They learn that geologists use tools to scratch the minerals, then examine the minerals with a hand lens to see whether marks have been left on the surface. The geologist suggests using a fingernail (a soft tool), a penny (a medium-hard tool), and the end of a paper clip (a harder tool).

Procedures

❖ What should be changed in the investigation?
   Four different minerals will be scratched.

❖ What should be kept the same?

Data Organization

How can we organize the data?

Notebook Entry (developed after small-group and class discussion)

Scratch Test

	Fingernail Yes or No	Penny Yes or No	Paperclip Yes or No	Total No. of Scratches
Mineral 1 Name				
Mineral 2 Name				
Mineral 3 Name				
Mineral 4 Name				

Result

The students learn that quartz (which can be white, pink, purple, or gray) is the hardest of the four minerals. Gypsum is the softest of the four. Calcite is softer than fluorite. Fluorite comes in a variety of colors, including white, green, blue, and violet.

A third grader produced the following work (and science notebook entry) while conducting this investigation.

Minerals and their Properties

	shape	color	texture	other
#1	no definite shape	green	bumpy and smooth	opaque
#2	rounded-square	orange	bumpy and smooth	opaque
#3	square	clear	bumpy and smooth	trans-lucent
#4	pentagon	brown	bumpy	opaque

(Minerals)

Scaffolding Science Inquiry Through Lesson Design

We can tell some properties by looking at them, like color, shape, texture. Some were green, orange, clear, brown. Some were shapes and some weren't any shapes, and some you can see through some. The chart does not help us decide hard or soft.

Scratch testing Minerals

Minerals	finger nail	Penny	paper clip	# of scratches
#1	no	yes	yes	2
#2	no	yes	yes	2
#3	no	no	no	0
#4	yes	yes	yes	3

#1 and #2 were the same. on the test #1 scratched #2. #2 did not scratch #2.

Order of Minerals

soft #4 Gypsum
    #2 Calcite
    #1 Fluorite
hard #3 Quartz

Procedures (class chart)

❖ What should be changed in the investigation?

The places where the battery, bulb, and wire are touched or how they are connected.

❖ What should be kept the same?

The three materials to be connected (one battery, one bulb, one wire).

❖ What should be observed or measured?

Whether the bulb lights or does not light.

Materials

What materials and tools will be needed?

Battery, bulb, wire

(Put names and pictures of materials on the word wall for reference.)

Data Organization

How will we organize the data/evidence in our notebook?

There are two categories of information

1. Ways of connecting the battery, bulb, and wire that made the bulb light.
2. Ways of connection the battery, bulb, and wire that did not make the bulb light.

Notebook Entry (most groups selected a T-chart)

Connections That Lit Bulb	Connections That Did Not Light Bulb
Diagram three or more connections that lit the bulb. (Draw clear and accurate diagrams.)	Diagram three or more connections that did not light the bulb. (Draw clear and accurate diagrams.)

Model for students a simple way to draw and label a clear and accurate diagram of a battery, bulb, and wire:

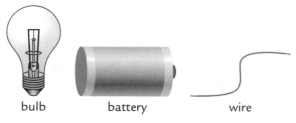

bulb   battery   wire

A fourth grader produced the following work (and science notebook entry) while conducting this investigation.

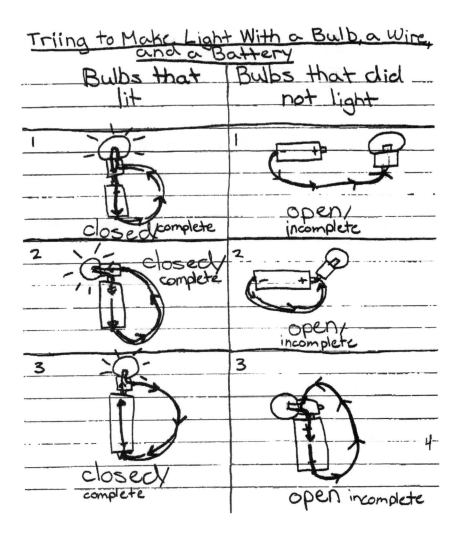

Triing to Make Light With a Bulb, a Wire, and a Battery

Bulbs that lit	Bulbs that did not light
1 closed/complete	1 open/ incomplete
2 closed/ complete	2 open/ incomplete
3 closed/ complete	3 open incomplete

* As I added 1 more bulb
they both got dimmer.

bright

did not light — was not connected to the bottom

* both bulbs
were dim

6

Scaffolding Science Inquiry Through Lesson Design

Pathway Through the Bulb

(bulb)
energy receiver

glass
bead
– keeps
wires
apart

filament
thin – gets
hot because
it's like having
to go single file
– lights

support
wire

side
terminal

base
terminal

thin = more
resistance

Pathway

energy
source
(battery)

8

Procedures (class chart)

❖ What should be changed in the investigation?
  The methods used to separate the mixtures.

❖ What should be kept the same?
  The mixture of water, gravel, diatomaceous earth, and salt are to remain the same. One level teaspoon of each solid material will be placed in 50 ml of water.

❖ What should be observed or measured?
  Recovery of the solid in each of the mixtures and a comparison of each mineral's properties before and after being recovered from a mixture.

Materials (listed on the word wall)

Gravel, salt, diatomaceous earth, screen, coffee filter, beaker, filter apparatus, stirring stick.

Data Organization

How will we organize the data/evidence in our notebooks?
(Small-group discussions followed by a general class discussion.)

Notebook Entry

(Most small groups devised a data organizer similar to the one below.)

Mixing and Separating Mixtures

	Observation After Stirring	Screen: What Happened?	Coffee Filter: What Happened?
Gravel			
Diatomaceous earth			
Salt			

A fifth grader produced the following work (and science notebook entry) while conducting this investigation.

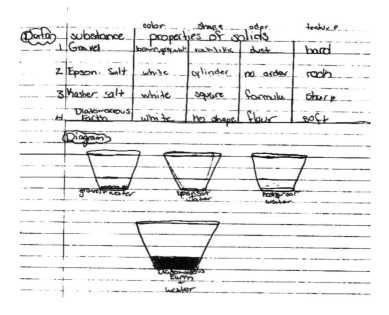

Data | substance | properties of solids
		color	shape	odor	texture
1	Gravel	brown, gray, whit	rock like	dust	hard
2	Epson. Salt	white	cylinder	no order	rock
3	Kosher salt	white	square	formula	sharp
4	Diatomaceous Earth	white	no shape	flour	soft

Diagram

Separation Strategy

Solid substance	screen	coffe filter	other
gravel mixed w/water	yes	no	
Powder mixed w/water	no	yes	
salt crystals mixed into water	no	no	no

Procedures

❖ What should be changed in the investigation?

❖ What should be kept the same?

❖ What should be observed or measured?

Materials

What materials will we need to conduct the investigation?

Sequential Plan for Conducting the Investigation (Notebook Entry)

First, _____ .

Second, _____ .

Then, _____ .

Finally, _____ .

Data Organization

❖ What kind of graphic organizer can be used to record the data collected?
❖ What will a reasonable data organizer look like?

Notebook Entry

Using this graphic organizer, record observations as the investigation is carried out.

Making Meaning

7

After students have recorded their focus questions, predictions, plans, and data organizers in their science notebooks, conducted their investigation, and organized and recorded the data they have gathered, they need to interpret and clarify what they have learned by (1) participating in a classroom discussion; (2) writing claims and evidence; (3) drawing conclusions; and (4) engaging in reflection (see Figure 7–1).

Explanation fosters a deeper understanding of science. Preparing evidence-based scientific explanations of their investigations helps students understand natural phenomena, articulate that understanding, and convince others that their understanding is accurate.

Helping students develop evidence-based explanations is challenging. They typically discount data that contradict the predicted

Figure 7–1 Sequential Tasks in Implementing Curriculum

Phase 1: Set the Stage for Learning
Vocabulary Development
- ❖ kit inventory
- ❖ dynamic word wall

> Discussed in Chapter 4

Phase 2: Formulate Investigable Questions and Predictions
- ❖ engaging scenario
- ❖ focus question (notebook entry)
- ❖ prediction (notebook entry)

> Discussed in Chapter 5

Phase 3: Plan and Conduct the Investigation
- ❖ sequential plan (notebook entry)
- ❖ data collection methods and forms (notebook entry)
- ❖ data obtained (notebook entry)

> Discussed in Chapter 6

Phase 4: Make Meaning
- ❖ discussion/analysis (conference)
- ❖ claims and evidence (notebook entry)
- ❖ conclusions (notebook entry)
- ❖ reflection (notebook entry)

> Discussed in this chapter

outcome of their investigation or make assertions that have little backing or justification in the data they have collected. They have difficulty using appropriate evidence and connecting that evidence to a claim.

Teachers need to make the underlying framework of scientific explanation—claims, evidence, and conclusions—explicit. A *claim* is a deduction, a pattern, or a discovery garnered from the investigation. *Evidence* is the data observed and measured during the investigation. The claim is linked with the evidence in the *conclusion.*

Metacognitive strategies help students understand their unique learning patterns. One definition of *metacognition* is knowing what

Scaffolding Science Inquiry Through Lesson Design

we know and what we don't know. It may also be thought of as the ability to:

1. Devise a strategy for acquiring needed information or solving a problem.

2. Be aware of the steps that comprise the strategy.

3. Reflect on and evaluate the productivity of the strategy.

Some metacognitive strategies (skills) that may help students understand their own thinking and thereby increase their ability to learn science include forming questions and predictions (see Chapter 5) and creating plans and graphic organizers (see Chapter 6).

Class discussions and science notebooks are integral metacognitive strategies that require the communication skills of listening, reflecting, and being concise. Having students rephrase what they have just said and asking them to consider other students' ideas as well as their own are ways of building on and extending ideas. Teaching students about their own thinking is as important as exposing them to the content of science.

Writing is an important tool for making coherent, structured claims based on evidence, and students need strong scaffolds in order to do this well (Klentschy and Molina-De La Torre 2004; Songer 2003; Hug et al. 2005). The three guiding principles for science instruction recommended by the National Research Council (2005) are:

1. In order to activate their prior knowledge, students must be engaged.

2. In order to develop competence in an area of inquiry, students must:

    a. Have a deep foundation of factual knowledge.

    b. Understand these facts in the context of conceptual frameworks.

    c. Organize knowledge in ways that facilitate retrieval and application.

3. Metacognitive approaches to instruction can help students take control of their own learning by defining their goals and monitoring their progress.

Science instruction in which students learn definitions, label preprinted diagrams, and complete worksheets does not build student understanding. Reading science texts, keeping a science notebook, and conducting investigations are, in themselves, insufficient as well. Instruction that leads to an understanding of science's big ideas is developmental: individual lessons are the building blocks of subconcepts, and subconcepts are the building blocks of big ideas.

Discussion/Analysis (Conference/ Class Data Organizer)

Having students plan and conduct investigations does not in itself constitute a high-quality science program. If instruction stops there, it simply uses activities for activities' sake. After students conduct their investigations is when most science is learned. Students need to make sense of the observations they have made and the data they have collected during the investigation. This may be best accomplished during a class discussion, or *making meaning conference*, in which the class uses a data organizer to record and display the data and observations each small group has collected during the investigation. The class then determines the patterns and relationships revealed in the collected data by noticing replications or variations. These measured or observed patterns and relationships become the basis (evidence) for claims (explanations).

Figure 7–2 is a class data organizer for a fourth-grade investigation of static electricity. After the class compiled the results of each group's investigations on this chart, the classroom teacher used the following questions to prompt students to look for patterns and construct a set of claims supported by these patterns.

- What do the data tell us about how static electricity works?

- Let's analyze the similarities between static electricity and magnets.

- What claims can we make about plastics and paper?

Scaffolding Science Inquiry Through Lesson Design

Figure 7–2 Class Data Organizer for Static Electricity Investigation

Items	Trial I		Trial II	
	Attracts Paper	Did Not Attract Paper	Attracts Paper	Did Not Attract Paper
Grocery bag				
Sandwich bag				
Balloon				

To help the students connect their claims with the appropriate evidence, the teacher provided the following scaffold.

Claims	Evidence
I claim that . . .	I know this because . . .

Two students shared the following claims and evidence:

Claims	Evidence
I claim that . . .	I know this because . . .
I claim that static electricity in an object is stationary and that when directed to an object with an opposite charge, it gets attracted to it.	I know this because when we rubbed the plastic against the floor it got negatively charged; then when we placed it over the paper it became attracted to it because the paper was positively charged.
I claim that in a magnet the opposite poles (N and S) attract and that in static electricity, positive and negative attract.	I know this because we saw the N and S poles of the magnet attract, and we saw the paper attract the plastic bag after the plastic bag was dragged across the carpet.

When several other students commented that they had similar findings, these claims and supporting evidence were accepted. The discussion continued until all claims that could be supported by evidence were listed on the class chart.

Claims and Evidence (Notebook Entry)

Good claims (explanations) are based on evidence gathered during investigations (how we know) (National Research Council 1996, 125). Although diagrams and pictures can provide some evidence of conceptual understanding, a recounting of procedures does not constitute a claim. Students need to write about what they learned, not what they did. Students must be guided to look at the patterns (evidence) in the data (what they observed or measured) and then make claims (explanations) supported by this evidence. The most significant conceptual change occurs when students take a position regarding an investigation and provide evidence for that position: that is, when they write about *what they have learned from* the investigation rather than *what they did during* the investigation. Students are now using their observations to support their reasoning.

Conclusions (Notebook Entry): Tying It All Together

Now it's time to go back to the intended curriculum's guiding questions. The following prompts help students draw conclusions.

- How do your claims and evidence relate to the big idea?
- How were your predictions supported by our evidence?
- How would you revise or change your thinking based on the evidence?
- What did you learn that was new?

Students then complete a science notebook entry (Today I learned . . .) stating their conclusion(s).

Reflection (Notebook Entry)

The final step in implementing curriculum is for students to write a reflection in their science notebook. This metacognitive strategy is an important part of making meaning and can be encouraged by prompts such as:

- What other things did you wonder about?
- Are there any new questions you have about your investigation or next steps you want to take?

These prompts should change from lesson to lesson. Their purpose is to get students to focus on what they learned, not what they liked or didn't like about the investigation or how they felt about working with other members of the group.

Students can then share their responses and perhaps create an Interesting Questions class chart that includes suggestions for the best way to get information (e.g., interview an expert in person, on the telephone, or via email; search the Internet; read a book or an article;

conduct an experiment). Students could also pursue one of these questions for homework or as an extension or enrichment activity.

Making Meaning Examples

The remainder of this chapter provides three examples of the making meaning phase of implementing curriculum. After studying these examples, you can use the template in Figure 7–4 to help you prepare for your making meaning conferences.

Making Meaning from the Earth Materials Lesson

Class Data Organizer Related to Determining Hardness of Minerals

Properties of Minerals—Hardness Scratch Test

Tools Used	Fingernail (Yes or No)	Penny (Yes or No)	Paperclip (Yes or No)	Total No. of Scratches
Mineral 1 (Name)				
Mineral 2 (Name)				
Mineral 3 (Name)				
Mineral 4 (Name)				

Focus Questions

1. How can we use the results of the scratch test to rank the minerals from softest to hardest?

2. What can we claim about the hardness of each of these minerals?

3. What is our evidence? (Evidence relates back to the planning question: What [the scratches] are we going to observe or measure?)

4. The geologist at the museum said that quartz is the hardest of the four minerals, that gypsum is the softest, and that calcite is softer than fluorite. How can we use the results of the scratch test and what the geologist told us to name the minerals?

Scaffolding Science Inquiry Through Lesson Design

Organizer for Claims and Evidence

Softest → Hardest

Softest
What mineral do you claim is the softest? What is your evidence?

Next
What mineral do you claim is the next softest? What is your evidence?

Then
What mineral do you claim is next to the hardest?

Finally
Finally, what mineral do you claim is the hardest? What is your evidence?

Claims and Evidence (Notebook Entry)

Conclusions (Discussion and Notebook Entry)

1. How do the claims and evidence relate to the big idea?
   (Big idea: Rocks and minerals have different properties that are determined by how they are formed and that make them useful in different ways.)

2. What use do people make of minerals?

3. What minerals are in or around your house? (Homework assignment)

4. How did the evidence support your prediction? How would you revise your thinking based on the evidence?

5. What did you learn that was new?

6. Complete this sentence in your science notebook: Today I learned . . . (Share several examples with the class.)

7. Write any new questions or next steps for your investigation in your science notebook. (Share several with the class.)

Reflection (Notebook Entry)

Quick-write: What was the most important point you learned in this lesson? What new questions do you now have?

In this plan, the teacher has constructed a class data chart on which students from each working group will post their results. He has also developed some focus questions to stimulate class discussion and reiterate the content goal: to perform a scratch test to determine the hardness of four minerals and then rank them from hardest to softest. The teacher repeatedly asks students for evidence to support their claims and asks them to write their own claims and evidence statements in their science notebook. He then asks students to support or revise their original prediction based on the new information they've gathered and complete the Today I learned . . . statement in their science notebook. The lesson concludes with students writing a reflection in their science notebook.

In the example that follows, the student restates the focus question (what he wanted to find out) and makes two claims supported by evidence from the investigation. Because no conclusion has been stated nor a reflection completed, this is a work in progress.

Focus Question

How can we describe the minerals? How can we put them in order from soft to hard?

Claim	Evidence
Some thing about minerals can be described by looking at them.	By looking you can describe color, shape, texture, and if light will go through them. Colors were green, orange, clear, brown. Most were bumpy with smooth places. One mineral was translucent.

continue on next page.

Scaffolding Science Inquiry Through Lesson Design

Claim	Evidence
Scratching the minerals with a penny, finger nail and paperclip can decide the hardness	Gypsum - softest It had marks from all 3 tools. Calcite- It had marks from 2 tools, and from fluorite. Fluorite- It had marks from 2 tools and was not scratched by Calcite. Quartz- hardest It had no marks.

In the next example, from a lesson later in this unit, the student, in a series of discursive paragraphs, makes a clear set of claims and evidence statements regarding the properties of igneous rock and expresses a clear conclusion.

claims and Evidence:
I claim that rocks that como from Volcanoes are hard and shiny/glassy. When we checked Rock #9, it was black, hard, and shiny. Rock #9 come frome a volcano. Its name is obsidian. It is a type of igneous rock.

> I learned that some rock come from volcanoes. If they come from volcanoes, then they are called igneous rock Two examples that are igneous rock are obsidian and pumice. I learned that pumie rock has tiny holes. Those holes are pockets of air.

Another student linked the evidence from the investigation to his claims to draw a conclusion regarding what he learned.

> What we learned— I learned that some minerals are harder than others. For example, mineral A be sraghed by the fingernail, penny, and nail but mineral E could only be scratched by the nail The harder the mineral the less streak powder would be left on the plate. Mineral E had a big black streak, and it seem to be a softer mineral. Mineral A also was soft and left a lot more powder than others

Making Meaning from the Magnetism and Electricity Lesson

In this example, the teacher asks students to make a small chart showing ways the circuit either lit or did not light.

Scaffolding Science Inquiry Through Lesson Design

Class Data Organizer

Referring to the data in their notebooks and using strips of chart paper folded into four sections, half the groups will share an example of a circuit that lit, the other half will share an example that did not light.

From your notebook, copy a diagram of a connection that lit the bulb.	Diagram just the energy source (the battery). Show the contact points with large dots.	Diagram just the energy receiver (the bulb). Show the contact points with large dots.	Why do you think the bulb lit? Explain.

From your notebook, copy a diagram of a connection that did *not* light the bulb.	Diagram just the energy source (the battery). Show the contact points with large dots.	Diagram just the energy receiver (the bulb). Show the contact points with large dots.	Why do you think the bulb did not light? Explain.

Reminder: Evidence relates back to the question What are we going to observe or measure?

Claims and Evidence (Notebook Entry)

Conclusions (Discussion and Notebook Entry)

1. How do the claims and evidence relate to the big idea?
   (Big idea: Electrical energy can be converted into heat, light, sound, motion, and a magnetic field.)

2. How were your predictions supported by the evidence?

continues

3. How would you revise/change your thinking based on the evidence?

4. What did you learn that was new?

5. How is electrical energy used to produce light, sound, motion, and heat in your home? What effect would an open or closed circuit have on your examples? What might cause a circuit to be open? (Homework assignment)

6. Complete this sentence in your science notebook: Today I learned . . . (Share several examples with the class.)

7. Write any new questions or next steps for your investigation in your science notebook. (Share several with the class.)

Reflection (Notebook Entry)

Quick-write: How were your predictions supported by the evidence? How would you revise/change your thinking based on the evidence?

One student in this class listed the following claims and evidence, conclusion, and reflection.

Claims&Evidence. I claim that the metal tip and the metal are the critical contact points. I know this because we placed the wires on different places and the didn't worked. When we place them on the critical pants they lighted up.

I claim that a circuit is a path of electricity. I know this because we connected the wires to the battery and the other side to the light bulb and the energy flow through.

Conclusion. The evidence in the experiment did not support my prediction because I connected the wires to the posetive side of the wires to the metal of the light bulb and it didn't worked.

Reflection. I would like to learn about circuit boards.

Another fourth-grade teacher who taught this lesson had her class write a discursive summary of the making meaning conference (see Figure 7–3). The opening (topic) sentence restated the focus question, the following sentences were claims and evidence statements, and the concluding sentence restated the topic sentence to reflect whether the prediction had been reaffirmed or revised based on what was learned during the investigation.

Organizer for Claims and Evidence

Topic Sentence: Focus question turned into a statement.	

Guiding Question 1 (from the intended curriculum)

Using a wire, a battery, and a bulb, explain what is required to light the bulb. What is the role of the battery, the bulb, and the wire?

Claim ⟶	**Evidence** (what we are observing or measuring)
Claim ⟶	**Evidence**
Claim ⟶	**Evidence**
Claim ⟶	**Evidence**

Guiding Question 2 (from the intended curriculum)

What is the difference between a closed and an open circuit?

Claim ⟶	**Evidence**

Conclusion: Restatement of the topic sentence.

The number of claims and evidence required to convey conceptual understanding will vary, as will the completeness and quality of the writing. Encourage students to use scientific language from the dynamic word wall.

An example of student work from this fourth-grade class is shown next.

Claims and Evidence

We can make light with a wire, a battery, and a lightbulb.

The critical contact points are the base and side terminals on the bulb and the negative and positive ends of the battery. The bulb did not light if any of these points were not touched.

Evidence

Example #1:
This circuit works.

Example #2:
This curcuit did not work.

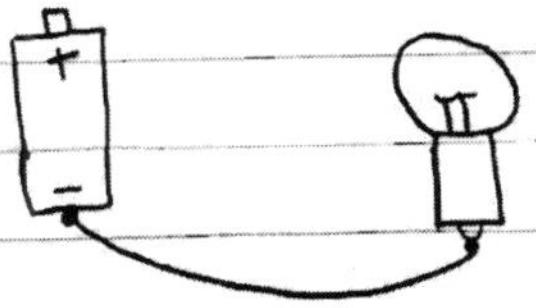

A bulb lights if have a complete loop, an energy receiver, an energy source and wires to make connections.

My prediction shows an incomplete curcuit. It would not light the bulb.

This student has rephrased the focus question (How can we make light with a battery, a wire, and a bulb?) as a statement (We can make light with a battery, a wire, and a bulb). Guiding question 1 is answered in the next sentence, with two illustrations included as evidence. Guiding question 2 is answered next. There is no conclusion, but the student does state that the prediction was incorrect. Asking Why didn't the bulb light? would prompt the student to offer a more detailed explanation.

Making Meaning from the Mixtures and Solutions Lesson

Class Data Organizer

Mixing and Separating Mixtures

	Observation After Mixing and Stirring	Screen: What Happened?	Filter Paper: What Happened?
Gravel (1 level teaspoon) Water (50 ml)			
Diatomaceous earth (1 level teaspoon) Water (50 ml)			
Salt (1 level teaspoon) Water (50 ml)			

Organizer for Claims and Evidence Discussion

Gravel

How separated? How are a screen and a filter alike and different?

Claim ──────────▶ Evidence

Before and after being mixed comparisons—How alike? How different?

Claim ──────────▶ Evidence

Diatomaceous Earth

How separated?

Claim ──────────▶ Evidence

Before and after being mixed comparisons—How alike? How different?

Claim ──────────▶ Evidence

Salt

How separated?

Claim ──────────▶ Evidence

Before and after being mixed comparisons—How alike? How different?

Claim ──────────▶ Evidence

Claims and Evidence (Notebook Entry)

Conclusions (Discussion and Notebook Entry)

1. How do the claims and evidence relate to the big idea?
   (Big idea: Substances have characteristic properties. A mixture of substances can often be separated into the original substances based on the properties of the substances.)

2. How were your predictions supported by the evidence?

3. How would you revise/change your thinking based on the evidence?

4. What did you learn that was new?

5. Give examples of mixtures not already mentioned in class that you might find in a kitchen, a restaurant, or your home. Explain why they are mixtures. (Homework assignment)

6. Complete this sentence in your science notebook: Today I learned . . . (Share several examples with the class.)

7. Write any new questions or next steps for your investigation in your science notebook. (Share several with the class.)

Reflection (Notebook Entry)
Quick-write on any of the discussion topics.

In the following example of student work provided, the student wrote a claim linked to each of the mixtures.

Scaffolding Science Inquiry Through Lesson Design

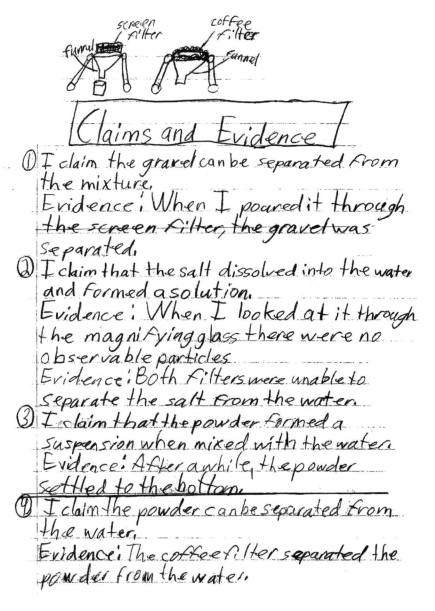

Claims and Evidence

① I claim the gravel can be separated from the mixture.
Evidence: When I poured it through the screen filter, the gravel was separated.

② I claim that the salt dissolved into the water and formed a solution.
Evidence: When I looked at it through the magnifying glass there were no observable particles
Evidence: Both filters were unable to separate the salt from the water.

③ I claim that the powder formed a suspension when mixed with the water.
Evidence: After awhile, the powder settled to the bottom.

④ I claim the powder can be separated from the water.
Evidence: The coffee filter separated the powder from the water.

In the additional example that follows, the student again responded to all the focus questions and also revised his prediction with new information learned from the lesson.

Claims	Evidence
solids can be seperated by a Screen.	I know this because solids are to big to fit in a screen only liquids can fit threw a screen.
some solids dissolve with water.	I know this because water makes it old so it dies out and it dissolves.
gravel can be seperated by water with a Screen.	I know because gravel is to big to fit in a Screen and all liquids can.
Diatomaceous Earth can be separated with a coffee filter and not a screen.	I know this because we tried it and the Powder turnes a little wattery that it does fit in the screen and since the coffee filter holes are small the Powder doesnt fit.
Mixtures did not seperated by filtering	because the Powder and gravel wer separed from the water.

Conclusion

My prediction was partially correct because my prediction for gravel and Powder were correct but my prediction for the mixtures salts were wrong because I dint include that they wer going to dissolve and turn from mixtures to solutions.

A final example, a third grade student's work, is provided to show how the student made meaning from a unit of study on brine shrimp.

What I Learned about Brine Shrimp:
- Brine Shrimp are crustaceans like sowbug, lobster and crabs.
- They have exoskeleton which is a hard outer covering like a shell
- They are small-less than 3 centimeters
- Brine Shrimp live in salt water of parks, lakes, and along the ocean shore.
- The eat tiny, tiny animals. We fed ours yeast.
- Brine Shrimp molt. When they outgrow there "shell" it splits open and falls off. A new, bigger shell is underneath.
- Brine Shrimp have many legs called appendages which can regrow.
- The famale brine shrimp has an egg sac the first time she has eggs.
- Brine Shrimp use their legs to swim. They swim upside down!

My favorite thing about brine shrimp is when they swim backwards.

Figure 7–4   Making Meaning Conference Template

Class Data Organizer

Organizer for Writing Claims and Evidence

Claims and Evidence (Notebook Entry)

Conclusions (Discussion and Notebook Entry)

Reflection (Notebook Entry)

Verifying the Achieved Curriculum

8

T he final component of the planning template for scaffolded guided inquiry is the achieved curriculum. Student science notebooks are the focal point of this phase (see Figure 8–1): What has been achieved in a lesson and how it aligns with the intended and implemented curriculums can be determined by examining students' science notebook entries. As a formative assessment tool, science notebooks provide valuable diagnostic evidence about what was learned. Science notebook entries reveal gaps between what was expected of each student in the intended and implemented curriculum and where the student is. Similarly, classwide results inform teachers about the effectiveness of the lesson's learning strategies and allow gaps in the instruction to be pinpointed and adjusted. To improve student learning, teachers must reflect on their own practice as well as examine the resulting student work.

What Constitutes Effective Feedback?

We have found that student communication skills and understanding do not improve by the mere use of science notebooks. Teachers also need to give students feedback. However, not all forms of feedback are equally effective.

Using a feedback tool such as a simple checklist to derive a grade or numeric score does not in itself provide the type of guidance and support required to improve students' learning. Scaffolded guided inquiry identifies areas in which the student can benefit from written feedback. When teachers provide comments that are nonjudgmental, are specific to the content, and indicate how improvements may be made, student interest and achievement are enhanced. (However, giving marks or grades *along with* comments can be just as ineffective as giving grades alone. Students focus on the competitive nature of the grade and often ignore the comments.)

Providing students with feedback on their work is one of a teacher's most important strategies, especially in inquiry-based science instruction geared toward *transforming* knowledge rather than *presenting* it. Marzano et al. (2001) reported, in a review of nine research studies, that providing feedback that guides students, rather than telling them what is right or wrong on a test, can improve students' standardized test scores by 30 percentage points. Therefore, the most appropriate form of feedback in a transforming approach to instruction is asking guiding questions (or writing them in the student's science notebook). For example,

> What evidence do you have in your data to support your claims?
>
> What claims can you make from your evidence?
>
> What is another explanation for what happened?

Teachers can also conduct a written conversation with the student in her or his science notebook.

Effective written feedback is content specific and includes:

- Evaluating the quality of the communication, including but not limited to the use of language standards and conventions and other criteria appropriate to the genre. For example, if the student

Scaffolding Science Inquiry Through Lesson Design

Figure 8–1   Achieved Curriculum

Intended Curriculum
Big Ideas—Public Announcement

Lesson Content Goals

1. ⟷
2. ⟷

Guiding Questions

1. Make public
2.

State Standard Addressed

Implemented Curriculum
Opportunities to Learn

Kit inventory
Dynamic word wall—synonyms (tags)
Engaging scenario—connect to world

Achieved Curriculum
Feedback Guide

Science notebook
Formative assessment
   of teaching and learning
Proficiency/guidance for
   improvement

❖ Focus Question —————————▶ ❖ Focus Question
   A question that leads to construction
   of knowledge about lesson content goals

❖ Prediction —————————▶ ❖ Prediction
   I think or predict that _____ because _____.
   If _____, then _____.

❖ Plan —————————▶ ❖ Plan
❖ Data Organizer   ❖ Data Organizer
❖ Data   ❖ Data
   Plan, organize, set expectations

Making Meaning Conference
• Class graphic organizer (key
  concept), thinking map
• Sharing data, group analysis
• Claims and evidence emerge—identify
  on organizer

❖ Claims and Evidence —————▶ ❖ Claims and Evidence
❖ Conclusions   ❖ Conclusions

Closure
• Share, discuss, challenge claims and
  evidence, revisit big ideas
• Revisit predictions —————————▶ ❖ Reflection
• Next steps, new questions   Support or change
                                        thinking

has written a prediction, does the prediction answer the focus question to the best of the student's current knowledge and contain a *because* clause that reveals prior knowledge and possible misconceptions?

- Pointing out and modeling how conceptual and procedural understanding is demonstrated through examples, comparisons, relationships, and the functional use of scientific language (science process skills).

- Recognizing the ability to comply with instructions within the context of individual creativity.

- Identifying and sharing exemplary work (the "wow" factor).

Possible starters for written feedback based on Bloom's taxonomy (see Bloom 1981) are shown in Figure 8–2.

The timing of feedback is also critical to its effectiveness. In general, the longer students have to wait to receive feedback, the longer it takes them to clear up misconceptions (Marzano et al. 2001). Besides being timely, feedback should reflect students' developmentally appropriate level of skill or knowledge (both expected and actual). Knowledge acquisition takes time and practice. Students do not remain engaged in learning if corrective and timely feedback is not given and success remains a mystery. Science notebooks should receive on-the-spot oral feedback while each lesson is in progress and detailed written feedback when the lesson is over. In addition, students need opportunities to respond to written feedback.

Research consistently indicates that this form of feedback has a more powerful effect on student learning than simply telling students their standing in relation to their peers (Marzano et al. 2001). Developmental "story lines" become a kind of graphic organizer or flowchart indicating the big idea or unifying concept of each unit of study (Amaral et al. 2002) and then the basis for a rubric against which to provide appropriate written feedback. (Figure 8–3 is a story line for a unit on magnetism and electricity.)

Scaffolding Science Inquiry Through Lesson Design

Figure 8–2   Feedback Starters Based on Bloom's Taxonomy

Cognitive Domain	Generic Starters
Knowledge	Who, what, where, when, why, how much, which one, describe, select, label.
Comprehension	What is fact? Opinion? Why? What does _____ mean?
Application	Explain what is happening. Substitute scientific words for _____. Give an example of _____ . What would happen if _____. Explain the effects of _____. Describe what might change if _____.
Analysis	What patterns/relationships do you see on the chart, graph, diagram, data chart? How are _____ and _____ alike/different? What idea/property is most important?
Synthesis	What conclusion can be drawn from _____? Create a _____ that shows _____ . Design a _____. Predict _____. Plan _____.
Evaluation	Which is most important? How would _____ change your mind about _____? Compare _____ to _____. Rank _____. How was/was not _____ a fair test? What would you change? How do you agree or disagree with _____?

Figure 8.3 Magnetism and Electricity Story Line

Unifying Concepts and Processes: Systems, order, and organization; evidence, models, and explanation; constancy, change, and measurement; equilibrium, form, and function.

Big Idea: Electricity and magnetism are part of a single force. This force has been used to advance technological discoveries.

Subconcept 1: Magnets attract and repel each other; iron objects stick to magnets.

Subconcept 2: An electric circuit is a complete pathway through which current travels.

Investigation 1: The Force

Part 1: Investigating Magnets and Materials	Part 2: Investigating More Magnetic Properties	Part 3: Breaking the Force	Part 4: Detecting the Force of Magnetism
Investigating magnetic force; observing which objects stick to magnets	Exploring induced magnetism	Measuring the force of magnetic attraction	Detecting magnetic force in a closed "mystery box"
NC 3.01	NC 3.02, 3.04	NC 3.01	

Investigation 2: Making Connections

Part 1: Lighting a Bulb	Part 2: Making a Motor Run	Part 3: Finding Insulators and Conductors	Part 4: Investigating Mystery Circuits
Exploring simple electric circuits	Building a motor circuit with a switch	Testing for conductors and nonconductors	Finding hidden circuits with mystery boards
NC 3.03, 3.05	NC 3.03, 3.06	NC 3.06	

Subconcept 3: Components can be added to a circuit as long as the complete pathway remains.

Subconcept 4: Electromagnets can be created by current flowing through a conductor.

Subconcept 5: Technology is using science knowledge to solve problems or improve existing objects.

Investigation 3: Advanced Connections

Part 1: Building Series Circuits	Part 2: Building Parallel Circuits	Part 3: Solving the String of Lights Problem
Making a series circuit from a simple circuit	Splitting a series circuit to create parallel pathways	Using knowledge of series and parallel circuits to solve problems
NC 3.07	NC 3.07	

Investigation 4: Current Attractions

Part 1: Building an Electromagnet	Part 2: Changing Number of Winds	Part 3: Investigating More Electromagnets
Creating an electromagnet with wire wrapped around an iron core	Experimenting with how the no. of winds of wire affects electromagnetic strength	Changing variables to change electromagnetic strength.
NC 3.02, 3.08	NC 3.02, 3.08	NC 3.02, 3.08

Investigation 5: Click It

Part 1: Reinventing the Telegraph	Part 2: Sending Messages Long Distance	Part 3: Choosing Your Own Investigation
Applying previous knowledge to build a working telegraph	Connecting two telegraphs in order to communicate	Planning and carrying out an investigation from an original question

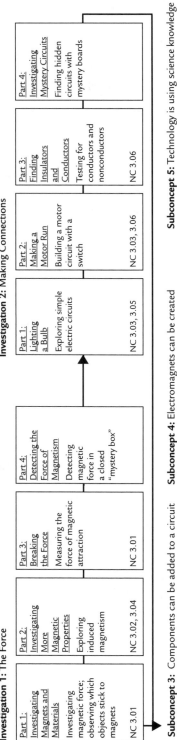

National Science Standards: Science as Inquiry, Physical Science, Science and Technology, History of Science

North Carolina Science Standards: Grade 4, Competency Goal 3, 3.01–3.08

Student Self-Assessment Based on Feedback Guides

Research (Marzano et al. 2001) indicates that given an appropriate guide, students can effectively monitor their own progress. However, effective self-assessment comes about only with the teacher's sustained and consistent guidance. Clearly stating the required learning tasks and goals gives students a way to judge where they are in relation to the goals, puts them "in the know" regarding their own learning, and is an important aid in their ability to construct meaning.

If teaching and learning assessments are to be fair and consistent, student science notebook entries should be measured against task-specific feedback guides that have been developed for each lesson and are specifically aligned with the entries required by the intended and implemented curriculums. They should be written in student-friendly language and shared with students at the beginning of each lesson. Making the learning goals for the lesson visible motivates and supports student success, and referring to the guides frequently during instruction helps students maintain their focus. They become self-regulated learners.

Figures 8–4, 8–5, and 8–6 illustrate feedback guides that establish expectations for what students need to accomplish in order to be proficient—quality of communication and conceptual/procedural understanding—and that students can use to evaluate their own work.

Figure 8–4   Scratch Test Feedback Guide (Earth Materials Unit)

Notebook Entry	Quality of Communication	Procedural and Conceptual Understanding
Focus Question	❖ Relates to scenario and content goals ❖ Cannot be answered yes or no ❖ Can be investigated with available materials ❖ Uses appropriate syntax and conventions—language is understandable	
Prediction	❖ Answers the focus question ❖ Uses *because* ❖ Is understandable	❖ Reveal prior knowledge and possible misconceptions
Planning	❖ Was done as a class and was not required to be copied into the notebook ❖ Data organizer(s) are titled, labeled, and appropriate to task	
Data	Organizer 1: Observable Properties of Four Minerals ❖ Information is accurate, complete, and understandable Organizer 2: Mineral Scratch Test ❖ Information is accurate, complete, and understandable	❖ Show care in conducting the test

Figure 8–4   Continued

Notebook Entry	Quality of Communication	Procedural and Conceptual Understanding

Claims and Evidence Statements	❖ Connect to the content and guiding questions of the intended curriculum
	❖ Are clear, understandable, and complete
		❖ All claims are accurately supported by evidence in student notebooks or class charts
		❖ Use scientific terms from the word wall with understanding
		❖ Show conceptual understanding of:
		  1. How minerals can be tested for the property of hardness
		  2. Correct ranking of the four minerals based on evidence

Reflection (Quick-write)	❖ Is understandable	❖ Accurately conveys what was learned in the lesson and why it is important

Notebook Entry	Quality of Communication	Procedural and Conceptual Understanding
Focus Question	❖ Relates to scenario and content goals ❖ Cannot be answered yes or no ❖ Can be investigated with available materials ❖ Uses appropriate syntax and conventions (language is understandable)	
Prediction	❖ Answers the focus question ❖ Uses *because* ❖ Is understandable	❖ Reveal prior knowledge and possible misconceptions
Planning	❖ Was done as a class and was not required to be copied into the notebook ❖ Data organizer(s) are titled, labeled, and appropriate to task	
Data	❖ Three accurate, clear, and labeled diagrams of bulbs that lit ❖ Three accurate, clear, and labeled diagrams of bulbs that did not light	❖ The complete loops of the diagrams of bulbs that lit are accurately traced in red with direction arrows

Scaffolding Science Inquiry Through Lesson Design

Figure 8–5 *Continued*

Notebook Entry	Quality of Communication	Procedural and Conceptual Understanding
Claims and Evidence Statements	❖ Connect to the content and guiding questions of the intended curriculum ❖ Are clear, understandable, and complete ❖ Are written as a titled complete expository paragraph of at least five sentences	❖ All claims are accurately supported by evidence in student notebooks or class charts ❖ Use scientific terms from the word wall with understanding ❖ Show conceptual understanding of a complete circuit, critical contact points, the role of the battery, bulb, and wire(s)
Reflection (Quick-write)	❖ Is understandable	❖ Accurately gives examples of electrical energy that produces heat, light, sound, and motion ❖ Accurately explains the effect of a closed or open circuit on the examples

Figure 8–6   Separating Mixtures Feedback Guide

Notebook Entry	Quality of Communication	Procedural and Conceptual Understanding
Focus Question	❖ Relates to scenario and content goals ❖ Cannot be answered yes or no ❖ Can be investigated with available materials ❖ Uses appropriate syntax and conventions (language is understandable)	
Prediction	❖ Answers the focus question ❖ Uses *because* ❖ Is understandable	❖ Reveal prior knowledge and possible misconceptions
Planning	❖ Data organizer(s) are titled, labeled, and appropriate to task	❖ Accurately identifies what is to be changed, what is to be kept the same, and what is going to be observed or measured
Data	Organizer 1: Properties of Substances Before and After Separating from Water ❖ Information is accurate, complete, and understandable Organizer 2: Separation Strategies ❖ Information is accurate, complete, and understandable	❖ Show care in conducting the procedures and tests

Scaffolding Science Inquiry Through Lesson Design

Figure 8–6 *Continued*

Notebook Entry	Quality of Communication	Procedural and Conceptual Understanding
Claims and Evidence Statements	❖ Connect to the content and guiding questions of the intended curriculum ❖ Are clear, understandable, and complete	❖ All claims are accurately supported by evidence found in student notebooks or class charts ❖ Use scientific terms from the word wall with understanding ❖ Show conceptual understanding of: 1. What constitutes a mixture (gives accurate examples) 2. How a mixture of gravel and water, a mixture of diatomaceous earth and water, and a mixture of salt and water can be separated 3. What properties affect separating a mixture of gravel and water, a mixture of diatomaceous earth and water, and a mixture of salt and water
Reflection (Quick-write)	❖ Uses clear and understandable language and conventions	❖ Accurately compares the prediction to what was learned in the lesson and how thinking would be revised based on new evidence

9

Documenting Effectiveness

The premise behind the scaffolded guided inquiry teaching model presented in this book is that aligning the intended curriculum (what should be taught—that is, the standards), the implemented curriculum (what is taught), and the achieved curriculum (what is learned by students)—the model proposed by Marzano (2003)—when combined with scaffold-supported science notebooks and class discussion, as described by Klentschy (2005), increases students' ability to arrive at evidence-based explanations of the results of their investigations and ultimately deepens their understanding of science.

In 2003, the California Mathematics/Science Partnership in Imperial County, California, began a collaboration with researchers from the Tennessee State University Center of Excellence for Learning Sciences' Academic Achievement and Teacher Development in Science (AATDS) project to examine the effectiveness of this approach. The research team has been conducting a series of longitudinal research investigations since the fall of 2004.

During the 2004–05 school year, three studies measuring student achievement were conducted in schools in Imperial County, California. Scaffolded guided inquiry lessons developed for the Full Option Science System (FOSS) unit on mixtures and solutions, a unit that reflects the content standards for fifth graders in California, were compared with existing FOSS units (which are kit-based) and with textbook-based instruction. (For a complete description of all three studies, see Vanosdall et al. 2007.)

The first study compared scaffolded guided inquiry using science kits with traditional textbook-based curriculums. Two districts in Imperial County were using textbook curriculums, one published by Harcourt Brace, the other by Macmillan. All four elementary schools in the two districts agreed to participate in the study. The final sample comprised twenty fifth-grade teachers and 563 students.

Ten randomly assigned teachers received the FOSS mixtures and solutions kits. (The California state content standards specifically assess the content covered in this six- to eight-week curriculum unit.) In place of the standard teachers guide accompanying the kit, the teachers received the scaffolded lessons guide and participated in six hours of professional development instruction on using the kit materials and presenting scaffolded lessons. The training took place in February 2005, and the teachers taught the unit during the spring semester of 2005. They received in-classroom support and coaching from resource teachers who were part of the Valle Imperial Project.

The ten remaining teachers followed their standard textbook curriculum and incorporated whatever "off-the-shelf" materials they typically used in their science classrooms. Their professional development consisted of the training offered by the respective publishing companies when the textbook series were adopted by the district. They too received in-classroom support and coaching from the Valle Imperial Project resource teachers.

Two measures of student achievement were used. One, the FOSS unit test (California edition), was specifically aligned with the mixtures and solutions unit. All the students were given one form of this test before instruction began and another form of it at the end of the unit. The other, which dealt with more general educational progress, was the California Standards Test (CST), a sixty-six-item standards-aligned test used for state accountability purposes. It is administered to all California fifth graders in the spring. Only the physical sciences subtest scores were used. The state does not set a formal proficiency standard for any subtest, but a score of 6 on each subscale (out of 11 on the physical science subtest) would lead to an overall score in the proficient range. Thus a score of 6 can be used as a rough guide.

Students in classrooms using traditional textbook-based instruction gained an average of less than 1 point on the FOSS posttest compared with the pretest. In contrast, students working with the science kits had an average gain of over 6 points. The mean posttest score of the science kit classes was over 5 points higher than that of the textbook-based classes. The mean score of the scaffolded inquiry classes on the CST physical sciences subtest was about 2.5 points higher than that of the textbook-based classes: just over 6 in the scaffolded inquiry group but less than 4 in the text-based group. Thus, the average student receiving scaffolded inquiry instruction would (roughly) fall into the proficient range, and the average student in the text-based instruction group would fall far short.

A surprising aspect of these results is that the gain in achievement of the scaffolded inquiry classes on the CST was somewhat larger than on the FOSS unit test, when one would have expected larger gains on the test linked specifically to the experimental curriculum. It could be because the intended curriculum for each FOSS lesson specifically targets at least one of the California state science standards, and the California state assessment, in turn, measures achievement of those standards.

A second study involved a group of teachers from three school districts that had participated in the Valle Imperial project since 1996. These teachers had used the FOSS materials a number of times and participated in professional development designed to deepen their own content understanding and address pedagogical issues.

Twenty-four fifth-grade teachers in eleven schools in the three districts were matched by background (years teaching in the system; number of hours of professional development; and experience with kit-based curriculum, in that order). One of each of the twelve matched pairs was then randomly assigned to the experimental group, the other to the control group. The total number of students in the classes was 762.

Teachers in the experimental group received the FOSS mixtures and solutions kit, the scaffolded lesson guide, and training in the use of this approach. Teachers in the contol group received only the FOSS kit. Both groups of teachers received the normal classroom support available to all Valle Imperial teachers. Training for the experimental group teachers took place in the fall of 2004, and the units were taught in either the second or third trimester of the 2004–05 school year.

The achievement tests administered were the same as those used in the first study. Again, all the students were given one form of the FOSS unit test before instruction began and another form of it at the end of the unit. The CST was administered in the spring, per state schedules and guidelines. Only the physical sciences subtest scores were used.

The mean FOSS unit pretest scores for the two groups were nearly identical (10.79 for the experimental group; 11.06 for the control group). Students in classrooms using only kit-based instruction gained an average of 1.6 points on the posttest. In contrast, students who also received scaffolded guided inquiry instruction gained an average of almost 6 points on the posttest. The mean FOSS unit test score of the scaffolded inquiry classes was about 4 points higher than that of the kit-only classes. The mean score of the scaffolded inquiry classes on the CST was 2.2 points higher than that of the kit-only classes. Once again, the scaffolded guided inquiry group outperformed the control group. It therefore appears that scaffolded lessons have an important and significant "additive" effect on kit-based instruction equivalent to one year's worth of growth (which of course includes the effects of maturation and whatever else is learned in and out of school).

The fact that scaffolded guided inquiry used in conjunction with kit-based materials leads to higher student achievement than traditional textbook-based instruction and that scaffolded guided inquiry in conjunction with kit-based materials leads to higher student achievement than instruction using kit-based materials alone raises the question of whether instruction using kit-based materials alone would lead to higher achievement than textbook-based instruction. We decided to try to answer this question.

In a follow-up study, the teachers in the experimental groups in both studies were then asked to use the scaffolded guided inquiry approach to teach another science unit. The teachers from the first study used traditional textbooks; the teachers from the second study used kit-based materials. The results of this study are necessarily more equivocal, because the teachers were not randomly assigned.

Assurance that the two groups were essentially identical except for the curriculum being used comes from pretest data and the fact that the teachers in both studies came from adjacent districts in a rather homogeneous rural area.

The data obtained in this third study demonstrated that scaffolded guided inquiry used in conjunction with kit-based materials dramatically improved fifth-grade science achievement when compared with either text-based instruction or instruction using kit-based materials alone. This suggests that although kit-based instruction in the hands of experienced teachers may be more effective than text-based instruction, the effects of scaffolded guided instruction appear to be much greater and far-reaching. The study also showed that teachers who have no prior experience using kit-based materials may be just as effective in using scaffolded guided inquiry with those materials as teachers who do. This suggests that the use of scaffolding to guide inquiry-based instruction may make it possible to implement inquiry-oriented reforms more rapidly and effectively.

Additional longitudinal studies were conducted in 2005–06 and 2006–07 in Imperial County, California. A replication study began in Wake County, North Carolina, in 2006–07. Preliminary data from these studies are consistent with the data obtained in 2004–05.

Thus a growing body of evidence indicates that scaffolding inquiry in connection with aligned curriculum that incorporates student science notebooks and class discussion holds great promise in providing all students with the opportunity to make meaning from their science investigations and develop the ability to arrive at evidence-based explanations of what they have learned. It is our hope that teachers will adopt these best practices in their own science classrooms.

Appendix 1

Blank Templates for Use in the Classroom

Blank Template

	Grade

Intended Curriculum

Big Idea (displayed in class during the lesson)

> **Establish working word wall throughout the lesson.**

>

> **Standard(s) addressed**

Lesson Content Goals	**Guiding Questions**
1.	1.

Implemented Curriculum
Engaging Scenario

Teacher's Notes

Focus Question (displayed, discussed by groups, recorded in notebooks)

Guide students to *how* or *what* questions.
You may wish to start with a class chart of group focus questions.

Prediction (discussed by groups, recorded in notebooks)

Prediction (optional format)
If _____ then, _____ because _____ .
I think that _____ because _____ .

Planning

General Plan

What are we going to keep the same?

What are we going to change?

What are we going to observe or measure?

Operational Plan

What will be the sequence of steps or procedures you will follow to conduct your investigation?

First, . . .

Second, . . .

Third, . . .

Next, . . .

Finally, . . .

Data Chart

Before you conduct your investigation, how will you collect your data?

What will your data collection device look like?

Making Meaning Conference (Teacher Directed)

1. Making a class data chart

After completing the charts in their notebooks, the groups share their results with the class. Record the data on the board or overhead transparency; discuss and analyze results. Students can add observations and other comments to their notebooks.

2. Looking for patterns from the data charts

Teacher asks: What does this tell us? Is there a rule? Is there a pattern here? (focusing on the lesson content goals)

Teacher will guide students in writing claims and evidence based on the data chart(s) as students share information. Teacher will assist students in associating their claim with the evidence. **Once finished, teacher states "Based on our claims, lets revisit the guiding questions and discuss the answers based on our evidence." Teacher then goes over the guiding questions and makes sure students are able to respond to them.**

Examples

Claim	Evidence
I claim that . . .	I claim that because . . .
I know that . . .	I know that because . . .

Claims and Evidence (record in notebook)

Instruction

Teacher says: "You are now going to write your own claims and evidence in complete sentence form. I want you to look at the focus questions and write a claim and evidence that answer these questions. Revisit the class data chart(s) to get your claims and evidence. Record them in your notebook."

Closure (record in notebooks)

Students are then asked to revisit their prediction and write a sentence that states whether the evidence from their charts supported it or not. They are to explain why. They are to provide clear explanations regarding how their evidence supported their prediction. Use the scaffold "Today I learned . . ." to assist students in stating what they learned from the investigation.

Reflection (record in notebooks)

Students will revisit the big idea; they may write new questions or design new experiments.

Achieved Curriculum

Proficiency Feedback

_____ **Focus question:** is the purpose, the problem that needs to be solved.

_____ **Prediction:** shows the relationship between the substances and process for separating the mixture.

_____ **Data:** there are _____ charts and illustrations to be considered. They must be accurate and complete. The drawing should be labeled.

_____ **Claims and evidence:** show evidence of understanding the lesson content goals, referring back to the guiding questions and focus question.

_____ **Conclusion:** shows evidence of an understanding of the lesson content goals.

_____ **Reflection:** next steps and new questions regarding the investigation, what else would the student like to know about this content, and logical connections to past lessons.

Appendix 2

Sample Lesson: "Mixtures and Solutions"

Intended Curriculum

Big Idea (display in class during the lesson)

Elements and their combinations account for all the varied types of matter in the world.

Establish a working word wall throughout the lesson.

Making and Separating Mixtures

Investigation 1.1 from Mixtures and Solutions

Standard(s) addressed

Lesson Content Goals	**Guiding Questions**
1. A mixture is a combination of two or more substances.	1. What is a mixture?
2. Some mixtures can be separated by filtering.	2. How can a mixture be separated?
3. Each substance in a mixture keeps its own properties.	3. What happens to the properties of substances in a mixture?

© 2008 by Michael Klentschy and Laurie Thompson, from *Scaffolding Science Inquiry Through Lesson Design.*
Portsmouth, NH: Heinemann.

Implemented Curriculum

Teacher's Notes

Advanced Preparation

Set up a material station for the gravel, diatomaceous powder, and salts. Put out a spoon and a craft stick for each group, and have one pitcher filled with water available for all to use.

This lesson will take at least three days to finish.

Materials for Groups of Four Students

1 spoon gravel	4 craft sticks (to stir)
1 spoon of diatomaceous powder	4 hand lenses (upon request)
½ liter container with water	8 sticky notes
1 spoon of kosher salt	1 filter
1 spoon of Epsom salt	1 funnel with stands
8 plastic cups	1 screen
1 syringe, 50 ml	
1 pitcher	

Save salt solutions for the next lesson.

Lesson Overview

Brief Planning

- Read engaging scenario.

- Discuss and share the problem to solve (focus question).

- Observe and list the properties of the solids; distribute the solid materials and hand lenses.

- Design a chart for properties of the solids; share information with class; and record it in notebooks.

- Distribute the syringes, cups, and spoons for students to mix a spoon of solids in 25 ml of water.

- Observe the mixtures and make diagrams in notebooks.

- Write a prediction in notebooks; students decide how the mixtures will be separated.

- Plan how to record the results on a chart.

- Discuss how to separate the mixtures.

- Test predictions; students separate the mixture with the screen first and discuss their results.

- Give instructions on how to use filter; students separate mixture with filter.

- Discuss the results and record information on charts.

- Making meaning conference; Guide making claims based on the evidence found.

- Claims and evidence; Students make claims and evidence statements focusing on guiding questions.

- Conclusions; Revisit prediction and write whether or not it was supported by the evidence.

- Reflection; Write a new investigable question to continue further investigation on the topic.

Day One

Engaging Scenario

Word wall synonym
Mixed: combined, put together

Read the engaging scenario; use the template at the end of this lesson to make either copies or a transparency so students can read along with you.

Teacher says:

Engaging Scenario

"Yesterday you and a couple of your friends were playing with some materials your dad was using for a project at work. You thought it might be fun to see what happens when you mix the materials in water. But before you did this, you thought it might be a good idea to weigh the substances first, using your dad's scale. Then you threw the solid substances into containers filled with water. The materials are now mixed in the water, and your dad wants you to return them to their containers. You and your friends have a problem. What is the problem you need to solve?"

Word wall
Solids: matter with definite shape and volume

Word wall synonym
Property: characteristic, a feature

Teacher's Notes

Word wall
Mixture: combines two or more materials that retain their own properties

Instructions

To help students understand the problem, they will first observe the substances.

Teacher says: "In order to solve this problem, I will show you the solid materials as they were before they were mixed in water. You will make a list of the characteristic properties of these substances, including the water. Can someone tell me an example of a characteristic property?" (color, shape, odor, texture) "What do we use to do observations?" (our senses) "We will also need to measure the amount of mass of each substance using the scale." Discuss with the group how they will record their observations.

Teacher says: "You will observe the gravel, the kosher salt, the Epsom salt, and the diatomaceous powder. You need to make a chart to list their characteristic properties." Allow students to discuss with their groups on how they will record their observations, then have them share some of them with the class. Do not start the activity until they are ready with their charts. (See chart one, following.)

Chart One

Properties of the Solids

	Substance	Color	Shape	Odor	Texture	Mass (g)
1	Gravel					
2	Kosher salt crystals					
3	Epsom salt crystals					
4	Diatomaceous powder					
5	25 ml of water					

Have students gather materials from the material station (1 spoon of each): gravel, kosher salt, Epsom salt, and diatomaceous powder in a labeled cup. Supervise as students record their observations and assist them when necessary. Have them measure and record the mass of each of the solid substances (#1–4). Allow some reasonable time for them to collect data. Then have students share the properties they observed.

Teacher says: "Because water is a liquid, measuring its mass is not straightforward, so it will have to be done in a separate step. We can't just put the water on the scale by itself because it will spill over. The water has to be placed inside one of the cups. How can we figure out the mass of the water by itself?"

(Allow time for the students to discuss. The answer is to measure the mass of the empty cup, then measure the mass of the cup with water, and then subtract. Use chart two below to organize.)

Also, point out to the students that liquids like water have no shape of their own—they take the shape of the container that holds them. Also, water has no color of its own—water is colorless.

Chart Two

Figuring the Mass of a Liquid (Water)

Mass of empty cup (g)	
Mass of cup with 25 ml water (g)	
Mass of 25 ml water by itself (g) (subtract)	

Teacher says: "Now that you have information on the substances, you will mix the solids with the water so you can observe the mixtures. This time you will make a drawing showing how each substance looks once you mix it with water. Make sure you label your pictures in detail. Now go get 25 ml of water for each of the cups."

Have the students mix the solids with the water and stir with the craft stick (see samples of illustrations below).

Data to Record in Notebooks

Illustrations of the substances in water after they are mixed.

Example

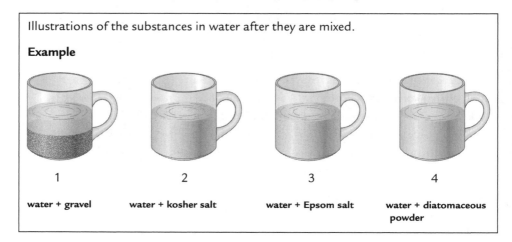

1	2	3	4
water + gravel	water + kosher salt	water + Epsom salt	water + diatomaceous powder

Teacher's Notes

After students have made the mixtures and recorded their observations, have a class discussion of what they wrote. Make sure everyone has more or less the same information on the charts. Use an overhead transparency to fill in the information they share as a class. Check that charts are filled correctly and provide feedback if necessary. Grade this notebook component as you check. Have students return the materials to the material station, these will be used the following day.

Note for Classroom Management

In order to do the calculation demonstrating the conservation of mass when substances are combined (lesson 5), it will be necessary for the students to get the same cups back the next day. To avoid any confusion, the cups need to be marked by group numbers and stored in such a way as to ensure that the students get back the same cups as the ones used when measuring their masses.

Day Two

- Return to the engaging scenario; reread the problem students need to solve.

- Solicit focus question samples from class and select the most appropriate one.

- Students write the class focus question in their notebooks if it is different from their own.

- Check notebooks and provide feedback.

- Remember to grade each notebook as you check them.

Focus Question (display, group discussion, and record in notebooks)

> **Suggested class focus question**
>
> How can solids mixed in water be separated?
>
> **Check and provide feedback.**

Teacher's Notes

Prediction (discuss in groups and record in notebooks)

Students will then discuss how they could solve the problem by writing a prediction. They work in groups discussing how and what they will use to separate the solids from the mixture. Remind students to use the kit inventory or word wall materials for ideas on possible equipment they can use to solve the problem.

 Provide prediction formats to help them write. They have to include a *because* statement to justify their prediction. Allow some time for discussion. Then have them share their ideas with the class. Predictions are recorded and checked as they write them.

> **Prediction: Examples (Optional Formats)**
>
> If we use the filter to separate the substances, then the water will come out and the substances will stay on the filter.
>
> *or*
>
> I think that if we use the screen to separate the substances, then the salt mixtures and diatomaceous powder will not separate because they will go through the screen.
>
> **Check and provide feedback.**

As the groups finish sharing predictions with the class, they are to decide on the way they will test and organize the data. Offer suggestions on how they can do this; look at chart two.

Then check/grade notebooks as they work. Have students share with the class ideas on how they will test their prediction. (See example below.) Do not start activity until the chart is ready.

Teacher says: "Keep in mind the following tips when you are testing. If you decide to use the filter only with all the substances, remember to change filters. Check that a cup is under the funnel to hold anything that goes through the filter or screen. Set up the funnel stand, fold the filter paper in half, then in half again, and open it like a cone cup. Place it inside the funnel. Make sure you know which substance you are testing by looking at the label outside the cup. If you are having some problems do not hesitate to ask for help."

Have the following materials available for the groups to test: funnel, stand, cups with mixtures, screen, four filters, four empty cups. Students may begin separating the substances and recording observations.

Chart Three

Separation Strategy			
Solid Substance	**Screen**	**Filter Paper**	**Other**
Gravel mixed into water			
Powder mixed into water			
Epsom salt crystals mixed into water			
Kosher salt crystals mixed into water			

Once students have finish testing, have them return the materials and clean their area. Save the salt solutions for the next lesson. Make sure the cups are marked by group numbers so that the students can get their own cups back in the following lesson.

Walk around the groups and listen to their discussions; make sure they are clearly recording their findings.

Day Three

Making Meaning Conference (Teacher directed)

1. Share and display class findings.

Remember, this is the conferencing stage; students are not required to include this in their notebooks. Have the class provide you with their findings. Create a chart on an overhead or poster to make the claims and evidence (use template at the end of this lesson).

2. Looking for patterns from the data charts

Guide students in writing claims from the evidence collected.

Teacher asks:

- "What claim can we make on the way gravel can be separated from water?"
- "What claim can we make about what happens to the properties of a substance when it is mixed in water?"
- "How can we determine if something has dissolved in water?"
- "What do the data tell us about how a powder can be separated from water?"
- "What is the rule or pattern on separating substances from water?" (Focus on the lesson content goals.)

Examples

Claims I claim that . . . I know that . . .	Evidence I claim this because . . . I know this because . . .
1. The gravel did not dissolve.	It was visible and it sank.
2. The salt mixed in water.	It disappeared as we mixed it.
3. Some of the powder mixed but not completely.	Some settled down at the bottom.
4. The salt did not separate with the sifter and the filter.	The water went through the sifter and nothing was left on the filter.

Teacher decides to list as many claims needed for students to understand the process.

Once finished, **teacher says:** "Based on our evidence, let's revisit the guiding questions and discuss the answers." Teacher then goes over the guiding questions and makes sure students are able to respond to them.

Claims and Evidence (record in notebooks)

Teacher can say:

"You are going to write your own claims and evidence statements. I want you to look at the guiding questions and write a claim using the evidence that answers these questions and record them in your notebook. Remember that we:
(1) combined substances together and created a mixture; (2) made claims based on the size and type of solids when mixed in water; (3) made claims on the process that was used to separate them; (4) discussed what happens to the properties of substances after you separate them. Write down at least three claims and evidence statements. Remember, these claims and evidence must be based on your data."

Example sentence structures:

I claim that _____. I claim this because _____.

I know that _____. I know this because _____.

Check and provide feedback.

Conclusion (record in notebooks)

Students are then asked to revisit their prediction and write a sentence that states whether the evidence from their charts supported it or not. They are to explain why.

Check and provide feedback.

Reflection (record in notebooks)

Students will revisit the big idea and their results. Students record an idea on how to separate the mixture of salts from the water.

Check and provide feedback.

Achieved Curriculum

Proficency Feedback

This guide is:

- to be posted on the board/butcher paper so students know what is expected in their notebooks

- to be an evaluation tool for teachers as students work in their notebooks

- to be converted into questions as a student self-assessment piece

Proficiency Feedback Guide **Mixtures and Solutions**	
	Focus questions • the problem relates to scenario

	Prediction
	   • shows the relationship between the substances and process for separating the mixture.
	   • uses *because*

	Data
	   • three charts, completed and with accurate notes
	   • drawings of mixtures, labeled

	Claims and evidence • three complete sentences showing understanding of content goals/guided questions
	Conclusion • accurately shows how prediction was supported or not
	Reflection • record statement on how to separate salts from water

References

Amaral, O., L. Garrison, and M. Duron-Flores. 2006. "Taking Inventory." *Science and Children* 43 (4): 30–33.

Amaral, O., L. Garrison, and M. Klentschy. 2002. "Helping English Learners Increase Achievement Through Inquiry-Based Science Instruction." *Bilingual Research Journal* 26 (2): 213–39.

Applebee, A. N. 1984. "Writing and Reasoning." *Review of Educational Research* 54: 577–96.

Aschbacher, P., and A. Alonzo. 2006. "Examining the Utility of Elementary Science Notebooks for Formative Assessment Purposes." *Educational Assessment* 11 (3–4): 179–203.

Aschbacher, P., and C. McPhee Baker. 2003. Incorporating Literacy into Hands-on Science Classes: Reflections in Student Work. Paper presented at the American Research Association Conference, April, Chicago, Illinois.

Bloom, B. 1980. *All Our Children Learning.* New York: McGraw-Hill.

Duron-Flores, M., and E. Macias. 2006. "English Language Development and the Science-Literacy Connection." In *Linking Science and Literacy in the K–8 Classroom*, edited by R. Douglas, M. Klentschy, and K. Worth. Alexandria, VA: NSTA Press.

Fellows, N. 1994. "A Window into Thinking: Using Student Writing to Understand Conceptual Change in Science Learning." *Journal of Research in Science Teaching* 31 (9): 985–1001.

Glynn, S., and D. Muth. 1994. "Reading and Writing to Learn Science: Achieving Scientific Literacy." *Journal of Research in Science Teaching* 31 (9): 1057–1073.

Harlen, W. 1988. *The Teaching of Science.* London: David Fulton.

———. 2001. *Primary Science: Taking the Plunge,* 2d ed. Portsmouth, NH: Heinemann.

Howard, V. A., and D. Barton. 1986. "Thinking on Paper: A Philosopher's Look at Writing." In *Varieties of Thinking: Essays from Harvard's Philosophy of Education Research Center,* edited by V. A. Howard. New York: Routledge, Chapman and Hall.

Hug, B., J. Krajcik, and R. Marx. 2005. "Using Innovative Learning Technologies to Promote Learning and Engagement in Urban Science Classrooms." *Urban Education* 40: 446–72.

Jorgenson, O., and R. Vanosdall. 2002. "The Death of Science? What Are We Risking in Our Rush Toward Standardized Testing and the Three Rs?" *Phi Delta Kappan* 83 (8): 601–605.

Klentschy, M. 2005. "Science Notebook Essentials." *Science and Children* 43 (3): 24–27.

Klentschy, M., and E. Molina-De La Torre. 2004. "Students' Science Notebooks and the Inquiry Process." In *Crossing Borders in Literacy and Science Instruction: Perspectives on Theory and Practice,* edited by W. Saul. Newark, DE: International Reading Association Press.

Marzano, R. 1991. "Fostering Thinking Across the Curriculum Through Knowledge Restructuring." *Journal of Reading* 34: 518–25.

———. 2003. *What Works in Schools: Translating Research into Action.* Alexandria, VA: Association for Supervision and Curriculum Development.

Marzano, R., D. Pickering, and J. Pollock. 2001. *Classroom Instruction That Works: Research-Based Strategies for Increasing Student*

Achievement. Alexandria, VA: Association for Supervision and Curriculum Development.

National Research Council. 1996. *National Science Education Standards.* Washington, DC: National Academy Press.

———. 1999. *How People Learn: Brain, Mind, Experience, and School,* edited by J. D. Bransford, A. Brown, and R. Cocking. Committee on Developments in the Science of Learning, Commission on Behavioral and Social Sciences in Education. Washington, DC: National Academies Press.

———. 2000. *Inquiry and the National Education Standards.* Washington, DC: National Academies Press.

———. 2005. *How Students Learn: History, Mathematics, and Science in the Classroom,* edited by M. S. Donovan and J. D. Bransford. Committee on How People Learn: A Targeted Report for Teachers, Division of Behavioral and Social Sciences and Education. Washington, DC: National Academies Press.

Reddy, M., P. Jacobs, C. McCrohon, and L. Herrenkohl. 1998. *Creating Scientific Communities in the Elementary Classroom.* Portsmouth, NH: Heinemann.

Rivard, L. 1994. "A Review of Writing to Learn in Science: Implications for Practice and Research." *Journal of Research in Science Teaching* 31 (9): 969–83.

Ruiz-Primo, A., M. Li, and R. Shavelson. 2002. Looking into Student Science Notebooks: What Do Teachers Do with Them? CRESST Technical Report 562. Los Angeles, CA: CRESST.

Scardamalia, M., and C. Bereiter. 1986. "Research on Written Composition." In *Handbook on Research on Teaching,* 3d ed., edited by M. C. Witrock. New York: Macmillan.

Shepardson, D. 1997. "Of Butterflies and Beetles: First Graders' Ways of Seeing and Talking About Insect Life Cycles." *Journal of Research in Science Teaching* 34 (9): 873–89.

Shepardson, D., and S. Britsch. 2001. "The Role of Children's Journals in Elementary School Science Activities." *Journal of Research in Science Teaching* 38 (1): 43–69.

Songer, N. 2003. Persistence of Inquiry: Evidence of Complex Reasoning Among Inner City Middle School Students. Paper presented at the Annual Meeting of the American Educational Research Association, April, San Diego, California.

Songer, N., and P. Ho. 2005. Guiding the "Explain": A Modified Learning Cycle Approach Toward Evidence on the Development of Scientific Explanations. Paper presented at the Annual Meeting of the American Education Research Association, April, Montreal, Canada.

Thier, M. 2002. *The New Science Literacy: Using Language Skills to Help Students Learn Science.* Portsmouth, NH: Heinemann.

Vanosdall, R., M. Klentschy, L. V. Hedges, and K. S. Weisbaum. 2007. A Randomized Study of the Effects of Scaffolded Guided-Inquiry Instruction on Student Achievement in Science. Paper presented at the Annual Meeting of the American Education Research Association, April, Chicago, Illinois.

Vygotsky, L. S. 1978. *Language and Thought.* Cambridge, MA: MIT Press.